PRAISE FOR *THE M3C MODEL OF COOPERATIVE CONTEXTUAL CHANGE*

The authors, Alessandro Biscaccianti, Mark Esposito, and Lloyd Williams, give us the keys of a simple and cooperative approach that allows all of us to succeed in strategies and approaches to change, personally and professionally. Use of this process allows one's fears, uncertainty, and routinized behavior—strong obstacles to change efforts—to be effectively embraced, reduced, and in many cases eliminated. Bravo to a new understanding and approach to change!

—Annie Bouvier, Director Human Resources, Merck Serono

THE M3C MODEL OF COOPERATIVE Contextual Change

THE **M3C MODEL** OF

COOPERATIVE Contextual Change

ALESSANDRO BISCACCIANTI, DBA • MARK ESPOSITO, PH.D.
LLOYD C. WILLIAMS, PH.D.

TATE PUBLISHING & *Enterprises*

Published by Tate Publishing & Enterprises, LLC
127 E. Trade Center Terrace | Mustang, Oklahoma 73064 USA
1.888.361.9473 | www.tatepublishing.com

Tate Publishing is committed to excellence in the publishing industry. The company reflects the philosophy established by the founders, based on Psalm 68:11,
"The Lord gave the word and great was the company of those who published it."

Cover design by Kellie Southerland
Interior design by Christina Hicks

Published in the United States of America

ISBN: 978-1-61777-191-0
1. Science / Social Aspects
2. Science / Applied Science
11.03.01

DEDICATION

To all those who believe in stability and change as foundations of life, love, happiness, and wisdom.

—Alessandro

To my eternal loves, Alina and Frappy; to my parents for their untiring support; and to all the people who constantly help me become who I am today. May change be your path to emotional and spiritual life enhancement.

—Mark

Dedicated to my sons, my grandson, my clients, and my students around the globe; may you continue to change cooperatively, contextually, emotionally, and psychologically as you impact the world.

—Lloyd

TABLE OF CONTENTS

PREFACE: THE M3C MODEL: FOCUSING CHANGE FORWARD

With over forty years of involvement in change initiatives ranging from the smallest unit to an organization of 10,000-plus, I suggest to fully appreciate the value of M3C, a retrospective consideration of the National Training Laboratory (NTL) and the Esalen Institute, as founders of change theory application, is beneficial.

In 1946, Kurt Lewin, as Director of MIT's Research Center for Group Dynamics, was focusing on the use of "training groups" to facilitate one's open-minded appreciation for and willingness to consider differences of values and perspectives. This underlying theory provided a foundation for the US Office of Naval Research and National Education Association to fund what became the NTL Institute for Applied Behavioral Science, which emphasized understanding the factors and conditions facilitating "changing behaviors." NTL's emphasis on developing "change agent" skills as well as "training group" methodologies for resolving confrontational differences, were and continue to be gradually absorbed into forward-looking business organizations. From this foundation but with a melding of diverse east/west perspective, the Esalen Institute was founded in 1962 by Michael Murphy and Richard Price, to explore the "human capacities," which enable dramatic change in the face of conditions, which are sub-optimal yet amenable if viewed from alternative perspectives.

The structural form for understanding and applying "change theory" offered by NTL and Esalen comprises: individual self-awareness giving rise to interpersonal competencies, evolving understanding of diverse perspectives through group dynamics, involvement-focused learning through sharing experiences, deconstructing the contextual components of the system within which the change will emerge, and building an orientation to growth by coupling the individual, group, and organization's aspirations.

Recent discourse on "change" reiterates the need for modeling that incorporates efficacy of functioning within complex, dynamic environments rife with uncertainty. For example, Baird and Griffin (2006) acknowledge that "at the organization level, we see a driving need for flexibility and adaptability to customers and the competition." Therefore, these authors emphasize the need for "dynamic learning," in that at any given time, some component of the organization is learning to be more efficient, effective, or qualitatively improved. From a developmental perspective, Worley and Lawler III (2006) offer a "build-for-change," the creating of an enterprise capable of orchestrating a variety of initiatives over time, which are responsive as well as anticipatory. In this regard, they stress that in a "world of perpetual change" an organization's leaders should adopt paradigms that facilitate regeneration.

Honoring the "structural form" of NTL and Esalen while embracing the context of "building for continuous qualitative improvement," M3C is capable of accepting various "change" models. M3C provides the foundational knowledge, the structural concepts, and the process-based actions for embracing change as a "forward learning" commitment. An especially pertinent takeaway from the M3C is the need to establish a sense of internal stability (the stake) among organizational members as an initial step for seeking "dynamic and sustainable" change. Because cooperation throughout the organization is both a "collective" and "individual" state of mind, the establishment of trust through respecting personal values and actively seeking input is essential for sustaining commitment, productivity, and qualitative performance. It seems appropriate to close with Schon's (1973, p.28) prescient statement,

> The loss of the stable state means that our society and all of its institutions are in continuous processes of transformation. We cannot expect new stable states that will endure for our own lifetimes.

The M3C acknowledges, accepts, and applies this dynamic reality.

—Thomas C. Neil, PHD

Thomas C. Neil, Ph.D. is a management consultant. Previously he was athletic director and chair business administration, Allen University (2004–06); chair of management, School of Business, Clark Atlanta University (1982–2003); the University of South Carolina (1977–1981); faculty University of Georgia (1970–1976). His primary interests are in the application of emotional intelligence as a communicative behavior, the relationship between emotional intelligence and knowledge acquisition and transfer in the learning process, the emotional component of creativity and innovation, and the use of human systems theory for improving organizational functioning. He received his PhD from the University of Florida in the areas of counseling, personnel, and Social psychology.

References

Baird, Lloyd and Darrell Griffin, "The Case for Dynamic Learning," Organizational Dynamics, 2006, Vol. 35, No. 4, pp. 372–383

Schon, Donald A., Beyond the Stable State. Public and private learning in a changing society, Harmondsworth: Penguin, 1973, 236 pages, p.28

Worley, Christopher G. and Edward E. Lawler III, "Designing Organizations That Are Built to Change," Human Resource Management and Industrial Relations, Leadership and Organizational Studies, October 1. 2006

Roots of "Change Theory" may be uncovered in the founding of the National Training Laboratory (NTL) and the Esalen Institute

EXECUTIVE SUMMARY

The book explores how a company can successfully alter its strategic policies and organizational methods in response to or anticipation of environmental changes without provoking instability or increasing the undue stress experienced by employees.

The model of cooperative contextual change (M3C) posits that to increase the effectiveness of the change process, it is necessary to introduce dynamic stability into the system (Senge, 1994; Desikachar, 1999). The premise of dynamic stability is that the whole organization and each individual should have a sense of security anchors represented by the following four elements:

1. To provide persistent direction and to make what is at stake explicit for everybody;
2. To promote co-building and co-opting of ground rules to handle the interfaces between the elements of the system and enforcement of these ground rules consistently;
3. To adopt systemic relationship methods to initiate and regulate actions at all levels throughout the company;
4. To foster a solution-oriented mindset.

The protocols used to implement the M3C and install the four elements of dynamic stability are proposed in this book. Of particular significance are the Contextual Drivership Method, which integrates leadership and management into organizational practices and processes, and the constructs of stake, reliability, and autonomy that assure the opportunity and capacity for success.

Finally, two integrative case studies describe the use of M3C in the turn-around of two industrial companies. The outcomes attributable to implementing the M3C were: increased coordination of people's actions, better capability to proactively anticipate issues, enhanced sharing of information by speaking a common managerial language, willingness to look for solutions instead of focusing on problems, and higher readiness to face future changes in a more cooperative way.

INTRODUCTION

In the contemporary economic context, change is unavoidable. Therefore, all stakeholders of a company—entrepreneurs, executives, managers, employees, and citizens—require considering the potential consequences change may have on the company.

In response to or in anticipation of the potential consequences, organizational leaders may introduce strategic and operational initiatives. These initiatives can be expensive, long-term actions, whose outcome is highly uncertain. In part, outcome uncertainty occurs because a "one best way," which ensures the response to change will be successful, is unavailable. In addition, imposed initiatives may generate a negative reaction instead of being accepted as a meaningful response to the environmental instability. Furthermore, although the conditions of change may provide an opportunity to leverage peoples' knowledge and skills, this benefit may go unrecognized (Haley, 1973, 1976).

When confronted with a new context and an imposed initiative, people may believe they have little vested interest because they don't own the change initiative (Watzlawick, Weakland, and Fisch, 1974). Failure of the change project may happen because people give lip service to the new behaviors while continuing to act in their old ways. Therefore, if the change initiative turns sour, the leaders and managers may decide to use their coercive power to achieve the desired results (Axelrod, 1984). Once coercion is introduced, the legitimacy of the change, as a benefit to all members, is diminished. The leadership is acknowledging that an acceptable rationale cannot be provided. Instead of being accepted as a source of stability—an effective response to the environmental conditions—the change initiative becomes a source of instability, generating conflict between the hierarchy imposing the initiative and those who must enact it.

The purpose of a hierarchy is to delineate the lines and levels of power and the authority relationships between individuals and groups. When the hierarchy requires actions perceived as legitimate, the legitimacy provides the sense of stability necessary to sustain daily operations, even in the face of change (Grant, 2001). Under conditions of widespread and/or intense change, stabilizing cultural, social, and/or psychological anchors become vulnerable. Therefore, sustaining existing anchors or providing alternative anchors is necessary if people are to focus their energy on positive actions (Watzlawick, Weakland, Fisch, 1974).

The typical business leader and manager believe change and stability are in contradiction; i.e., that change and stability cannot exist simultaneously. In fact, understanding that the paradoxical relationship, dynamic stability, can exist is an important lesson if contemporary business leaders and managers are to achieve successful change initiatives.

The model of cooperative contextual change (M3C) presented in this book describes the authors' understanding reached after years of management, consulting, and teaching experiences. This understanding evolved by discovering that a successful change process incorporated dynamic stability. The M3C is designed to create dynamic stability through the introduction, implementation, and sustaining of the following four anchors:

- Provide persistent direction and make what is at stake in the change explicit for everybody (Fisher, Ury and Patton, 1981, 1991)

- Promote the co-construction and the co-optation of the ground rules necessary for handling the interfaces between the elements of the system and enforcing the rules consistently (Axelrod, 1984; Grant, 2001);

- Adopt systemic relationship methods to initiate and regulate actions and make sure they are used at all hierarchical level throughout the company (Hersey and Blanchard, 1969, 1972, 1996);
- Foster a solution-oriented mindset (De Bono, 1985).

The M3C links several change process models and is designed to ensure that an organization, engaged in the process of change, achieves its goals. Its scope is sufficiently broad to allow its application to any type of organization (governmental, non-governmental, for profit, not-for-profit), regardless of the industry, the domain, or the size, for implementing change at any level (individual, team, department, company).

The application of the M3C is not predicated upon the existence of "change," but positive change is a consequence of its implementation. In fact, organizations that are experiencing success are encouraged to implement the model because it can further enhance success. The M3C establishes conditions that encourage co-operation, commitment, and the leveraging of resources that protect the stake of the organization. Because of these conditions, the M3C has face validity that encourages acceptance and implementation by any position in the company.

The model consists of four main modules: Contextual Assessment (CA), Community Norming (CN), Contextual Drivership (CD) and Cooperative Innovation (CI). Part three provides a full explanation of the protocols and tools useful to implement each module.

In particular, a new construct presented in this book is "drivership," an effective modality to provide organizations with coherent governance that offers a sense of stability, even as the process of change disrupts. Contextual drivership protocols are the result of the integration between the situational leadership and the accountability approaches with the principled negotiation approach developed at the program "On Negotiation" of Harvard School of Law (Fisher, Ury, and Patton, 1981, 1991; Lax and Sebenius, 1986). In the drivership paradigm, each individual concerned by the change process understands that enacting the driver role has a high symbolic value for the stability and the soundness of the relationship with all the other parties. Drivership provides an emotional anchor for the individual by clearly specifying the degree and level of responsibility by leveraging three main constructs: stake, reliability, and autonomy.

Two integrative case studies present the results of the implementation of the model of cooperative contextual change in two industrial companies. It is not possible to say that the outcomes observed are all directly related to the implementation of the M3C. However, the analysis of the managers' perceptions shows that they believe the use of M3C had a substantial impact on the cohesiveness of their team and helped them to leverage individual and collective competencies and resources to successfully carry out the turn around of their organizations.

Enjoy the text, as we believe its engagement can and will shift the dynamics and processes of your organization and its members to new heights!

—Dr. Alessandro Biscaccianti
—Dr. Mark Esposito
—Dr. Lloyd C. Williams

PART ONE: THEORETICAL FOUNDATION OF THE MODEL OF COOPERATIVE CONTEXTUAL CHANGE

1.1 ON CHANGE AND RESISTANCE TO CHANGE

1.1.1 Theories on Change

Beer and Nohria (2000) suggest two archetypes of change, theory E and theory O. For theory E, shareholder value is the only legitimate measure of corporate success. Change involves extensive use of economic incentives, drastic layoffs, downsizing, organizational re-engineering, and information technology. Change typically occurs through radical restructuring, with little consideration for the time or the process needed to change skills and behaviors. Radical restructuring is more likely to cause resistance and damage the positive elements of the organization's culture. In the long run, it often provides little return on investment (Anderson and Ackerman Anderson, 2001).

For theory O, the organization's ability to learn from its experience is the primary yardstick of success. The goal, employees committed to individual and organizational learning, should result in the improvement of organizational capability (Anderson and Ackerman Anderson, 2001). Achieving deep cultural change, from bottom-up improvement, is a long-term process.

Unfortunately, in the short term, the firm may generate insufficient financial performance to provide the resources necessary to sustain the learning initiative (Strebel, 1994, 2000). In theory E, the change initiator(s) forces change from the top disregarding the "human factor." In theory O, the change initiator(s) pushes responsibility to the bottom seeking buy-in, hoping this will lead to behavioral change.

Regardless of the selected change approach, the power to provoke, implement, and sustain change initiatives is the critical variable. Power emerges from the coalitions and alliances of those granted control of the resources and the authority to reward desirable behaviors (Schein, 1985; Kotter, 1995, 1996). The intersecting actions of the drivers of change and the forces of resistance generate four change patterns, shown in Table 1.1 (Strebel 1994, 2000). Strategic initiatives must be designed in response to the change pattern generated from the given context. Thus, "no change" resulting from strong resistance and weak drivers requires a different operational initiative than discontinuous change resulting from weak resistance and a strong driver.

		Drivers of change	
		Weak	Strong
Forces of resistance	Strong	no change	discontinuous change
	Weak	sporadic change	continuous change

Change patterns (Strebel, 1994, 2000)

Drivers of change may be external or internal and weak or strong to the organization (Anderson and Ackerman Anderson 2001). External drivers consist of the environment, marketplace requirements for success, strategic imperatives arising from industry characteristics, and governmental imperatives. Internal drivers are the organization's cultural imperatives, the interpersonal and intrapersonal behaviors of leaders and employees, and the mindset of leaders and employees. An individual using a theory E mindset may respond to external drivers by implementing a content/operational change that does not require leaders and employees to experience any deep, personal change. An individual with a theory O mindset seeks to use internal drivers to accomplish wide and deep organizational change to fit to the new content.

Executives, consultants, and academics' mental models, regarding change and/or their experiences with a theory, usually select an initiative that fits E or O. Beer and Nohria (2000) suggest that combining theory E and theory O is the most effective approach for creating dynamic, stable change in the short and long run.

Even if theory E and theory O are combined, when the change is externally induced, because the individual was not involved in selecting the change, he has no choice but to respond. This leads to fear, doubt, or anxiety, because the individual is required to give up habits and lose landmarks, resulting in instability (Beer and Nohria, 2000). When people are in a "contracted" state, they become emotionally preoccupied with their instability. As a result, their energies are spent fighting against the change instead of engaging in the behaviors necessary for successful change. Therefore, externally induced change often generates a failure circle, consisting of the conditions of stress and instability, as depicted in Diagram 1.1.

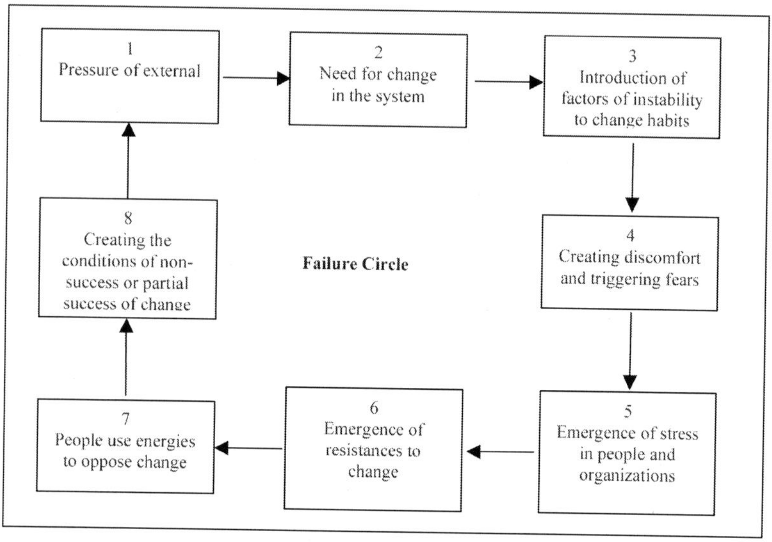

Diagram 1.1—The failure circle of instability-based change

The question is, "How do you jointly implement apparently contradictory theories?" when the possibility for inappropriate, overwhelming, or debilitating stress due to misunderstanding is high.

If an individual feels confident, calm, and centered in the face of change, he is more likely to engage in behaviors that support initiatives in response to change (Anderson and Ackerman Anderson, 2001). Therefore, change leaders must provide as much stability and emotional safety as possible (Schein, 2002). Leaders have to transform what is unknown into something that is uncertain in order to create order from the chaos and provide orientation when people face ambiguity (Schein, 1985).

As shown in the Table 1.2, two dimensions can be used to characterize the type of change and the type of design and implementation (Nadler and Tushman, 1986; Beckhard and Harris,

1987; Anderson and Ackerman Anderson, 2001; Meyer, Brooks and Goes, 1990).

Reactive change produces superficial results, which in the long term are unsuccessful. Therefore, the need for change evolves, finally reaching the transformational level, resulting in crisis management and a breakdown in the system. Reactive approaches to change are characterized by:

- Lack of a systemic approach: change is seen as a sum of single issues to be addressed rather than as a complex and evolving process;
- Top-down focus on strategic, organizational, or technological change with little consistency between the design and the implementation phases;
- Emphasis on short-term approach to increase efficiency (doing the least possible in the shortest time with as few resources as possible) without taking into account global performance (combination of efficiency and effectiveness);
- Lack of consideration of forces of resistance with little or no support for cultural, behavioral, and mindset change

Proactive change processes are considered the only viable approaches to the increasing complexity of the global business arena because they integrate all components of a system (Kotter, 1995, 1996; Pettigrew and Whipp, 1992; Anderson and Ackerman Anderson, 2001):

- Content refers to what the organization needs to change;

- Context refers to who has to change (the behaviors and state of mind of the human beings) as well as to the environment in which people live;

- Process refers to how the content and people changes will be carried out.

Developmental change is a prescribed enhancement of the old state, rather than a radical or experimental solution requiring profound altering of values, beliefs, and rules of the game. It is often a response to relatively small shifts in the marketplace, and implementation usually has a low risk of failure.

Transitional change is the response to significant shifts in the external factors perceived and experienced by the organization's decision-makers. During the transitional state, some component of the old state is replaced to achieve the desired new state.

Transformational change involves a radical shift from one state of being to another, which is an extremely painful process. Proactive transformation requires an awareness of the consequences the "new" context will have on the existing culture, behaviors, and mindset, if it is to be engaged in willingly.

Proactive transformational change requires that individuals align their thinking, their feelings, their intuitions, and their actions to permit the creation of a stable change process. Each individual must willingly witness her/his own experience, practice reflective thinking, be alert, clear-minded, observant, and open to any information coming from the environment.

		Type of change		
		Developmental	Transitional	Transformational
Design and Implementation	Proactive	Fine-Tuning	Evolution	Re-orientation
	Reactive	Adaptation	Re-organization	Crisis Management

Table 1.2—Types of change

The individual's willingness to commit is predicated on understanding the drivers of change, being able to recognize the triggering events, and proactively taking appropriate actions to start the transformation within self. This set of elements requires both a high level of conscious awareness (Anderson and Ackerman Anderson, 2001) and the recognition of one's individual stake in the change. Appendix One contrasts the three type of change with regard to ten main dimensions (Anderson and Ackerman Anderson, 2001):

McKinsey's 7-S Framework (Peters and Waterman, 1982), Nadler and Tushman's Congruence Model (1977), and Kotter's Eight-Stage Process of Creating Major Change (1995, 1996) offer organizing constructs for identifying what leaders need to pay attention to for change to have the desired impact. However, these models do not speak to creating conscious awareness. Therefore, the change driver is not provided with the conditions necessary for a dynamic and stable transformational change process (Anderson and Ackerman Anderson, 2001).

1.1.2 Systemic Approach to Change: The Communicative Therapy

The field of individual/family short therapy provides useful insights for enabling the emergence of conscious awareness and the implementation of effective change in business operations (Schein, 1985). Systemic family therapy recognizes that every individual is a vital part of a unit and cannot be treated in isolation. In addition, the aggregate unit must alter itself to sustain the new behaviors of each individual. From a systemic point of view, all systems tend toward a quasi-stable equilibrium. Therefore, any changing part of the system necessitates that all parts to a lesser or greater extent must change.

Theories developed to explain what happens to a business system over time usually focus on stability versus change, not on stability and change (Watzlawick, Weakland and Fisch, 1974; Haley, 1973, 1976). However, induced change must occur within a stable context if the desired results are to be achieved.

Table 1.3 shows the interdependence between the event of change and the type of condition (stability or instability) of the system during the change process:

Condition	No change	Change
Stability	Gradual Decline (backward-looking)	Evolution (forward-looking)
Instability	Disruption	Chaos

Table 1.3—Impact of stability and instability on change processes (Anderson and Ackerman Anderson, 2001)

A social system is stable when its members are able to be open, learn from their own and others' experience, discern what has to be done to protect the stake, give up what provokes problems, and detach themselves from contingencies by adopting reflective thinking. Stability implies durable absence of fear, discomfort, and conflict. Stable individuals are able to honestly assume their own responsibilities and act, regardless of the existing sources of trouble (Senge et al., 1994; Desikachar, 1999). The systemic approach developed by Erickson (1967, 1980, 1992) and by the therapists at the School of Palo Alto (Watzlawick, Beavin and Jackson, 1967; Watzlawick, Waekland, and Fisch, 1974; Watzlawick, 1983, 1984; Haley, 1973, 1976) identifies two different levels of change, first-order change and second-order change.

First-order change focuses on changing behaviors with no change of the frame of reference. The system then runs through many internal changes but doesn't improve and therefore declines, because the system holds the same old way of thinking.

Second-order change occurs when the system changes itself, which allows the system to evolve long-term. Second-order change requires the transformation of behavioral patterns to reflect paradigm shifts in the enactment of contexts. Every system and/or individual may experience second-order change, and when it happens, the system (or the individual) sees it as a discontinuity or a logical jump. From a first-order change point of view (that is from within the system), its practical manifestation is like a sudden illumination that often appears as illogical and paradoxical, as something uncontrollable, even incomprehensible.

Second-order change focuses on effects of future new behaviors and not on the presumed causes of existing behaviors. Although "why?" has always played a central role in the history of science, from the transformational perspective, the crucial question is, "What happens now?" "How to make it happen in the future?" and not, "Why did it happen?" (Watzlawick, Weakland and Fisch, 1974).

Often people try to resolve the issues at the content level; that is, by tackling the substance of the problem, even though the problem is at the relationship (meta-communicational) level (Watzlawick, Beavin, and Jackson, 1967). Change projects will be initiated around a new strategy, organizational structure, or quality process. Often, management gives the priority to the technical side of the change, focusing on like quantifiable issues, such as developing strategies, action plans, profit calculation, resources rationalization, while there is a tendency to ignore the human dimension (Arendt et al., 1995; New and Singer, 1983; Huston, 1992; Steier, 1989; Tessler, 1989; Levine, 1997). Little is then invested in the communication, training, and follow-up necessary to create the conditions for change to be successful (Schneider and Goldwasser, 1998) inputs that are initially required to people.

Most employees' needs are then ignored by management, creating disillusion, skepticism, and irritation (Weinbach, 1984). Manipulation and coercion are then used to overcome resistance, creating the conditions for the failure of the change process (Hultman, 1995; Carr; 1994; New and Singer, 1983). This type of technical approach tends to avoid or deny individuals' and teams' resistance to change. As a result, initially the behaviors and ways of relating will approximate the ways of acting associated with the new solution. However, because the players are only acting out roles, they will gradually regress to the old ways of behaving and relating (Schein, 1985).

Schneider, Ashworth, Higgs, and Carr's (1996) longitudinal survey demonstrates that is possible to change differently and more effectively. When changes in managerial practices, as observed by employees, precede changes in strategy implementation, change is more successful. This causal (or directional), positive relationship leads to an improvement in customer perception and intentions and, therefore, to higher organizational performance.

From Watzlawick, Beavin, and Jackson's (1967) perspective, problems arise when:

- Action is necessary but it is not taken (denying the problem);
- Action is taken when it is not necessary (the problem is either unsolvable or nonexistent);
- Action is taken at the wrong level (introduce first-order change when a second-order change is needed, or attempt second-order change when first-order would be enough).

Because second-order change requires behaving differently in relationship to an issue, the individual/system must re-frame the elements of the context. However, context re-framing is a difficult act, because the perceiver must step out of his point of view and adopt an alternative perspective. For this reason, the use of a change agent, someone capable of initiating an alternative perspective, and a facilitator, someone capable of sustaining the dynamic re-shaping, is crucial.

> It is not things themselves which trouble us, but the opinions we have about these things.
>
> —Epictetus

Communicative therapy posits that second-order change only happens when individuals modify the representation, the mental model they have of the reality. A mental model is a set of implicit and tacit assumptions and beliefs that drive the behaviors of the individuals (Senge et al.,1994). Transformational change begins with the individual making the mental model being used to frame the existing issue and its context explicit. Articulation requires the identification of the information selected from the context, explaining how the information is put together to interpret the perceived issue and what consequences are anticipated from the selected action.

A facilitator, using adroit reflective statements, helps the individual understand the enacted context, reframe it to match the desired stake, and, as a consequence, modify the associated behaviors (Schein, 1985).

Re-framing changes the meaning attributed to the situation and, therefore, the anticipated consequences. However, the facts associated with the situation are not altered. "To reframe means to change the conceptual and/or emotional setting or viewpoint in relation to which a situation is experienced, and to place it in another frame which fits the facts of the same concrete situation equally well or even better, and thereby changes its entire meaning" (Watzlawick, Weakland and Fisch, 1974).

Re-framing is an effective tool, because once individuals perceive the benefits of the alternative point of view, going back to the former view of reality outcomes is less desirable. Re-framing requires facilitators and the change agents to learn the manner in which the individuals communicate with each other. In addition, what the individuals present as "the way things are" must be accepted as the basis for change. The presented "this is the way things are" reveals the very resistance to change, which is then used to reference the re-framing and introduce change.

In the facilitated re-framing process, individuals analyze their role in the group and their interfaces with each other. Then they identify the issues created by such interdependence in their mutual relationships. Instead of focusing on each individual's responsibilities for the issues, the group focuses on the global functioning of the system and on how it could evolve. Each individual reframes his specific role and interfaces with the others instead of asking the others to change their behaviors. In this manner, the system is re-framed without individuals being asked to change explicit behavior (Schein, 1985).

1.1.3 Resistance to Change: A Resource to Make Change Happen

Resistance to induced change is expected yet difficult to assuage in a change process, because it can occur at three levels: individual, group, and organizational (Soparnot, 2003).

At the individual level, anxiety is the most important factor. The introduction of change breaks the existing equilibrium, resulting in uncertainty, which reveals itself as anxiety (Jacques, 1972; de Board, 1978). Whether the cause is real or imaginary, anxiety still produces the same psychological responses for an individual (Bovey and Hede, 2001b). The affective dimension, the perceived emotional relationship between the individual and the organization, is modified (Soparnot, 2003). The power relationship is altered because uncertainty is introduced into the zone of influence. Each individual perceives a divergence, which may enhance or diminish his interests (Crozier and Friedberg, 1977).

At the group level, the social contract under which the members lived is modified, altering the established norms. As a consequence, the members have to change their behaviors to comply with the new rules. These new rules impact the existing values of the group and may require the development of values, which could diverge from those held by individual members (Soparnot, 2003).

If the change is transformational, the identity of the organization may be modified and result in a new vision, mission, and stake. This new identity may generate defensive routines, subversive actions, and psychological as well as physical "absence" on the part of individuals as well as groups (Argyris, 1995; Soparnot, 2003).

Resistance is considered a threat to the efficiency and effectiveness of the change process at the strategic level (Maurer, 1996a). Yet resistance is a natural reaction designed to help individuals, groups, and organizations survive. The cognitive and affective role of resistance must be understood and leveraged to reduce irrational resistance (Bovey and Hede, 2001a; Maurer, 1996a; Waddel, 1998).

When facing change, individuals can direct energy to cope with or divert it away from the task of change (Oldham and Kleiner, 1990). When energy is diverted, it gives origin to maladaptive defense mechanisms (de Board, 1983). Bovey and Hede (2001b) consider resistance as the effect of defense mechanisms, which provokes behaviors that can be described in Table 1.4, resulting in the crossing of two continuums:

- the observability of the defense mechanism (from overt to covert)

- the impact of the defense mechanism on the individual's behavior (from active to passive)

	Overt (openly expressive behavior)	**Covert** (concealed behavior)
Active (originate action)	Oppose, argue, obstruct	Stall, dismantle, undermine
Passive (not acting, inert)	Observe, refrain, wait	Ignore, withdraw, avoid

Table 1.4—Mechanisms to resist change (Bovey and Hede, 2001b)

The same continuums can describe a taxonomy of the behaviors that individuals may have in accepting change, as shown by Table 1.5:

	Overt (openly expressive behavior)	**Covert** (concealed behavior)
Active (originate action)	Initiate, embrace	Support, cooperate
Passive (not acting, inert)	Agree, accept	Give in, comply

Table 1.5–Mechanisms to consent change (Bovey and Hede, 2001b)

Strebel (2000) recommends the use of a "personal contract" between the individual employee, the work group, and the change driver involving three levels of mutual commitments: a formal commitment on responsibility, a psychological commitment regarding involvement by each player, and a values commitment, which speaks to social identity.

Lipton (1996) and Gabriel (1992) argue that resistance experienced by the drivers of organizational change is similar to the resistance a psychotherapist faces when working with a patient. It is therefore legitimate to build a model of organizational change, drawing on a model of individual therapy (Olson, 1990).

Communicative therapy considers resistance a natural element of the system, the raw material that the therapist uses to understand the patient and as a source for involvement. Respect and listening for people who resist can effectively help to transform resistance into support (Maurer, 1996b).

The therapist leverages the patient's resistance by clarifying its function, which allows the patient to construct a personalized way to change (Erickson, 1967, 1980, 1992; Haley, 1973, 1976; Watzlawick, Weakland and Fisch, 1974; Watzlawick, 1983). Thus, for any change driver, the focus must be on creating the conditions that facilitate the willingness to become involved in change. The "new" paradigm of change is "learn how to change" rather than "manage the change" (Demers, 1999).

1.2 THE FOUNDATIONS OF M3C MODEL OF COOPERATIVE CONTEXTUAL CHANGE, AND ITS OPERATING CONCEPTS

1.2.1 On Contextualism

Contextualism posits that understanding the process of change, especially its consequences, can only occur when its specific context is recognized. (Pepper, 1942; Pettigrew, 1985a, 1987; Pettigrew and Whipp, 1992; Wilson, 1992). Contextualism views change as a continuous, iterative process, which is analyzed through different theoretical models (Brouwers et al., 1997).

Because change is a social construct, it requires the cooperation of participants who are willing to protect a common stake in spite of possibly divergent interests. If change is being considered as a means to improve an organization, its planning should encompass vertical and horizontal perspectives of the change initiative (Pettigrew, 1985b, 1987).

The vertical perspective articulates the relationship between the "environment" (what is beyond the control of the actors) and the "context" (what is within the control of the actors) (Hall and Fagen, 1956; Watzlawick, Beavin, and Jackson, 1967). The horizontal perspective focuses on the temporal connections between the interdependent actors and events as the community evolves from its present state to the desired future state.

"The human actor does not react to an environment, he enacts it" (Weick, 1979, p. 64). Actors' perceptions, their subjective interpretations, and mental representations of change shape both its context and process (Benson, 1977; Giddens, 1979, Pettigrew, 1990; Brouwers et al., 1997). Of particular significance are the enacted interpretations of the individuals who drive the change process (Whipp et al. 1988; Senge et. al. 1994).

To improve the change process, the way actors accomplish the change actions must be altered. Modifying the norms and rules of functioning prior to, during, and after change shapes the actors' perceptions regarding the actions necessary for accomplishing the change.

In the short term, modifying the norms and rules of functioning impacts the actors' behaviors. In the long term, it impacts their values. When modifying norms and rules, the history of the group of actors, their success and failures, and their relationships and conflicts must be considered (Weick, 1979; Giddens, 1979; Pettigrew, 1987; Brouwers et al., 1997). Any modifications need to be communicated using concrete examples of actions and successful outcomes.

1.2.2 On Cooperation

When groups focus on short-term gains as a reflection of narrow self-interests, social outcomes for both the groups, as well as the system, are often sub-optimal. Mutual defections from system policies and actions reflecting narrow self-interest typically occur when the individual or group does not perceive the benefits of and/or how to obtain mutual cooperation.

In a business context, the goals of change often reflect a suboptimal, zero-sum mentality, which results in winners at the expenses of losers, based on a struggle for resources and protection of individual advantages.

Mutual reciprocity, the recognition of interdependence, and the benefits of cooperative behaviors demonstrates that long-term solutions exist, which ensure payoffs to all without harming some (Axelrod, 1984; Wiley and Brooks, 2000).

Mutual reciprocity does not require trust between/among the actors, because defection is recognized as unproductive. Altruism is not needed. Reciprocity strategies can elicit cooperation even from an egoist. Finally, no central authority is needed because the acceptance of reciprocity generates self-policing. The willingness to engage in mutual reciprocity arises from an understanding that the achievement of self-interest in a social and economical environment is enhanced when long-term gains are maximized for all players. The emergence of mutual reciprocal cooperation fosters the collective use of shared knowledge to pursue common interests while minimizing coordination and control costs (Nalebuff and Brandeburger, 1996; Tsai, 2002).

Axelrod (1984) demonstrated that mutual cooperation among egoists is enhanced when they perceive a high probability of continually coming together to protect a common stake. The desired stake must be sufficiently important relative to the present stake. When the future stake is highly valued and other actors seek to cooperate in achieving it, the reluctant individual is more likely to engage in cooperative behaviors.

A change process based on mutual reciprocity requires involving the players in a learning process, which focuses on being able to articulate not only their self-interests and position but also the self-interest and position of the other players. The learning focuses on acquiring the skills of facilitative intercourse, extrapolation of probable consequences of alternative actions, and the design of interdependent behavior patterns. In this sense, the individuals are involved in learning the benefits of negotiation. The goal of the negotiation is the establishment of a contractual relationship, which links individual interests to protecting the common stake. This contractual relationship may be formal or informal. However, it must be based on co-built and co-opted ground rules and meta-communication protocols. In addition, all parties, to ensure drift has not occurred and the benefits of cooperation are still present, must continually review the premises of the contract.

The decision-making protocol that fits best with the constructs of cooperation and mutual reciprocity is consensus (Schutz, 1958, 1966, 1982, 1984). In consensus, a decision is made when all members agree with the decision, even if some disagree with the issue. Consensus, as a protocol for decision-making in a community, differs from unanimity (all members agree on the issue), from majority (more than half of the members agree on the issue), and also from authoritarianism (just one person agrees on the issue).

Consensual decision-making provides the community with more creative solutions, even if the process may last longer. Consensual decisions are usually executed more efficiently and effectively, since all members who will implement the decision have participated in the decision-making process. In fact, organizations are social communities that are capable of coordinating human behaviors better than market contracts because they provide "voluntaristic actions structured by organizing principles that are not reducible to individuals" (Kogut and Zander, 1992, p.384).

1.2.3 On Principled Negotiation

The effective enactment of change requires negotiation and the cooperation among all actors. Negotiation usually begins from a position of opposites. However, the Model of Principled Negotiation, developed at Harvard School of Law (Fisher, Ury and Patton, 1981, 1991; Ury, 1991), is a systemic approach, which focuses on having individuals understand the importance of making explicit what is mutually at stake.

In principled negotiation, individuals must initially be willing to identify and accept the mutual stake, i.e., that outcome which creates value for all. At that point, individuals may negotiate their individual stakes using complementarity as a guide. Individuals lacking formal authority within the organization need competence in negotiation to provide sufficient influence on the change process (Lax and Sebenius, 1986).

Operating Concepts of M3C: Stake

The stake is the most valuable tangible or intangible asset that an individual or a community wants to protect with their action.

The stake is the ultimate non-negotiable interest, the one that cannot be traded off. If the context changes, it may happen that the stake changes (and vice versa). The stake is the compass, the fixed reference that helps to define the direction and the orientation of the action: all actions must contribute to protect the stake. Clearly identifying the stake helps the drivers of the change to build and communicate the vision to be realized.

1.2.4 On Behaviors, Norms, Rules, Values

Constructivism holds that reality is a mental construction that does not exist independently of the inventor. Scientific, social, individual, and ideological realities are constructed because of an inherent human need to approach reality from a basic set of assumptions, which, although they are considered by their originators to be objectively derived, are in fact the result of the process of investigation. (Watzlawick, 1983).

If reality is not homogeneously perceived and the mission not well articulated, individuals will not have common objectives in the change process, which may cause a misinterpretation of the stake.

To achieve consensus requires the group to explicitly have a common language and share basic assumptions (Schein, 1985), which are translated into norms, values, and ground rules, resulting in the establishment of formal and/or informal contractual relationships among individuals, which are then bound together (Trice and Beyer, 1993).

Norms are the mutual sense a group has of what is right and wrong; they have a pragmatic dimension. Norms can develop on a formal level as written laws and on an informal level as social control and tradition. The norms, consciously or unconsciously, give people the feeling of security in knowing how to behave in "usual" situations. Norms provide the basis for organic solidarity through material interdependence and on division of labor (Durkheim, 1893, 1984; Morrison, 1995).

Values determine the definition of "good" and "bad." Values are closely related to the ideals people share and serve as a moral benchmark against which aspirations and behaviors are judged. Values provide the basis for mechanic solidarity through common beliefs, participation in collective action, and a common sense of identity (Durkheim, 1893, 1984; Morrison, 1995).

The lower the gap between values, norms, and behaviors, the higher the organizational performance and the lower organizational issues like high turnover (Kilman and Saxton, 1983; O"Reilly, Chatman and Caldwell, 1991).

A cultural way of thinking and doing evolves into a "reactive behavioral pattern" reflecting an existing norm, which is transmitted implicitly to all newcomers (Trompenaars and Hampden-Turner 1997). These implicit, cultural norms are largely unquestioned until they fail to produce desired results. At that point, the original stake must be presented, against which the reactive behavior can be compared.

Conventional approaches to team cohesiveness mostly deal with constructs related to the values of each individual (trust, respect, honesty, etc.). The assumption is that the organization will benefit because of the goodwill of the individuals to translate such values into actions.

Such an approach is affected by cultural bias, because it deals with a moral judgment, “What is good?” and “What is bad?” Working on “personal” values may result in a subjective and vague approach to the altering of interpersonal relationship, because different behaviors may originate from similar values and vice versa. Moreover, it does not provide the operational tools to prize virtuous behaviors and punish unwanted ones (Trompenaars and Hampden-Turner, 1997).

Alternatives approaches to team cohesiveness deal with the constructs related to the ground rules that are co-built and co-opted by the members of the team to protect the common stake and integrate knowledge (Grant, 2001). Co-building and co-opting ground rules generate a formal contract among individuals while protocols of communication and relationship constitute a meta-tool that functions as an informal contract.

At the cognitive level, ground rules shape the effort to learn how to relate with each other to deal with the external environment (solution finding). At the emotional level, ground rules clarify the issues related to authority and influence (member’s integration). Ground rules help the group to create solidarity and deal with conflicts (Schein, 1985). Co-building and co-opting reduces anxiety by helping the individuals accept each other, learn that they can live together, and know that cooperation is possible even if they do not all like each other. The emotional shift from “we all like each other” to “we all respect each other because we all respect the same rules” frees up emotional energy necessary to work together to achieve common goals. The establishment of mutual respect facilitates the leveraging of differences, encourages individual initiative, and reduces conformism.

Contracts are a way of life in universalist cultures. A contract serves to record an agreement on principle and codifies what the respective parties have promised to do. It also implies consent to the agreement and provides recourse if the parties do not fulfill their obligations (Watzlawick, Beavin, and Jackson, 1967; Watzlawick, Weakland, and Fisch, 1974).

Contractual relationships are seen as definitive by the universalist, but only as a rough guideline or approximation by the particularist. The latter will want to make the contract as vague as possible and may object to "restrictive" clauses. This resistance may indicate a preference for mutual accommodation rather than impending subterfuge. It is as well the sign of a higher importance given to the relationship, because relationships have a flexibility and durability that contracts often lack. Moreover, good relationships may involve a willingness to invest more (or differently) than the contract requires. Individuals should learn to distinguish interpersonal relationships from functional relationships in their day-to-day interactions to provide focused feedback.

Rule-based conduct (universalist) has a tendency to resist exceptions that might weaken the rules. The concern is that if one exception is allowed, the entire system will collapse. On the other end, particularist judgments focus on the exceptional nature of present circumstances. To reconcile this dilemma, ground rules are not considered absolute. Under certain conditions, ground rules or protocols may be changed through co-building and co-opting (Koestler, 1964; Trompenaars and Hampden-Turner, 1997). If people learn to work by integrating the universalist and particularist approaches, synergies are possible, and a virtuous circle can start (Trompenaars and Hampden-Turner, 1997).

1.2.5 On Leadership

Under the traditional paradigm of organizational hierarchy, leadership and management perform different functions and are associated with different stakes and behaviors.

Drucker (1954) lists five functions of the manager: setting objectives, organizing, motivating and communicating, measuring, and developing people. Mintzberg (1973, 1980) classifies the manager's work into three major functions (and ten roles): decision-making (entrepreneur, regulator, negotiator, distributor of resources), communicating (symbol, liaison, leader), and informational (active observer, transmitter, representative). Interestingly, Mintzberg considers leadership as one of the roles of a manager with soft skills being a major asset.

Bennis (1989) posited that leaders conquer the context while managers surrender to it. Table 1.6 contrasts the two roles.

Leader	Manager
Innovates	Administers
Develops	Maintains
Focuses on people	Focuses on systems and structure
Inspires trust	Relies on control
Long-range perspective	Short-range view
Asks what and why	Asks how and when
Has his eye on the horizon	Has his eye on bottom line
Challenges status quo	Accepts status quo
Does the right thing	Does things right

Table 1.6—Leader's features versus manager's features (Bennis, 1989)

The approaches distinguishing between transformational/charismatic, transactional, or passive leadership has contributed to perceiving the role of leader as multidimensional (Berson, Shamir, Avolio, Popper, 2002).

Transformational/charismatic leaders help their followers become more aware of the contribution of their tasks to the overall group's effort while transactional leaders tend to clarify long-term strategic goals and fulfill their followers' current needs (Bass, 1985).

Transformational/charismatic leaders appear to be more successful and effective, because they are able to move even the most resistant followers toward achieving a vision. These leaders rely on both inspirational and instrumental themes when they craft a vision statement. They link instrumental goals and objectives to the vision, creating a believable reality (Bass, 1960, Conger and Kanungo, 1998).

Table 1.7 shows how, under the traditional paradigm, leaders and managers interact with their subordinates. Leadership functions moved at the strategic level are more considered as detached from operations and more focused on a global role of visioning and mentoring. As workers and teams take on more managerial functions, managers are obliged to interact with them in a leadership capacity (Aubrey, 1999).

Leader
Power through influence
Looks at big picture, takes helicopter view of situation
Manages people as individuals
Helps people make their own rules

Table 1.7—Leader's and manager's interactions with subordinates (Aubrey, 1999)

An alternative perspective exists to the differentiation ("or logic") between leadership and management. This perspective ("and logic") recognizes that relational roles have more than one aspect. Value can be created for a role by adding complementary features (Watzalwick, Weakland, Fish, 1974; Nalebuff and Brandeburger, 1996). The "or logic" of leading and managing is seen as the origin of competitive and conflicting settings in the workplace (Axelrod,

1984). Contemporary global business environments compel the use of the "and logic" paradigm to build cooperative contexts.

Under the traditional paradigm, organizations establish a hierarchy consisting of leaders, managers, and workers. This hierarchy assigns to the leaders the responsibility for initiating actions in regard to change. It assigns to the managers the responsibility for regulating the implementation of the actions. It assigns to the workers the responsibility for doing the actions (Hersey, 1984; Bennis, 1989, Aubrey, 1999).

However, people in higher hierarchical position need to have access to their subordinates' knowledge to make sound decisions concerning strategic complex issues. If this is not possible, and lower levels of the organization are delegated only routine matters, the efficiency of the decision-making process is constrained by "bounded rationality" and "bounded access to relevant knowledge" (Grant, 2001). Therefore, under conditions of dynamic change, the leadership function has incorporated more of a global role of visioning and mentoring. Managers are obliged to accept more leadership responsibilities, and workers and teams must assume more managerial functions (Aubrey, 1999).

Within this re-structuring context, solutions are offered (empowerment, self-regulation, and team-centered decision-making), which are predicated on the existing traditional paradigm of hierarchy and leadership/management dichotomy.

These solutions ask both managers and workers to exhibit behaviors that combine leadership and management features. If initiating and regulating responsibilities are assigned to all levels within an organization while retaining the traditional hierarchal functioning, conflicts will arise.

Because these "solutions" contradict the existing paradigm, individuals become frustrated or anxious; the organization's leaders become disappointed with the lack of success and abandon the "solution" to look for the next newest solution. In this manner, the failure process begins again.

In the historical paradigm, the analysis of organizational authority and control within hierarchical organizations did not distinguish the role of managers in synchronizing the efforts of the specialists from the issue of promoting cooperation to ensure alignment in organizational goals (Grant, 2001).

The reason for this vicious cycle is that the historical paradigm introduces second-order change in the organization's procedures and/or processes while the constructs of hierarchy, leader, manager, and worker remain the same. Second-order change requires the changing of procedures and processes, which requires the reframing of the context (Watzlawick, Weakland and Fisch, 1974). Recognizing that relational roles may incorporate more than one dimension can alter leadership and management based on "differentiating logic" ("or logic"). The "or logic" of leading and managing is seen as the origin of competitive and conflict settings in the workplace (Axelrod, 1984).

Contemporary global business environments compel the building of cooperative contexts, which need to use the "and logic" paradigm. Value can be created for a role by adding complementary features (Watzalwick, Weakland, Fish, 1974; Nalebuff and Brandeburger, 1996).

The Situational Leadership approach (Hersey and Blanchard, 1969, 1972, 1974, 1996; Blanchard and Johnson, 1981; Hersey, 1984) and the Accountability approach (Swift, 2001; Gray et al, 1997; Woodward et al., 1996; Williamson, 1975; 1985) are among the operational models attempting to use an "and logic" to solve the leadership/management dichotomy.

Within a collaborative community, every individual is entitled to:

- Present what he believes is a deviation from the stake;
- Present a solution to the deviation;
- Propose an action that could impact the stake to the most appropriate strategic position.

Contextual drivership implies assigning each position within the organization the responsibility of protecting the stake. Incorporating into a position, at any level within the community, the responsibility of protecting the stake requires assigning the following duties: initiating actions to protect the stake and regulating stake-related activities. In fact, the quality of decision-making processes improves when the authority is delegated to those with the often tacit knowledge necessary to make sound decisions (Grant, 2001).

As a consequence, each position integrates leading and managing activities, which is a premise of the "and logic" approach. The M3C model requires modifications of the traditional "and logic" approach.

The purpose of organizational hierarchy must be limited to:

- Designating the strategic and operational positional activities, which support the organization's stake and coordinate the integration of the efforts of the specialists (Grant, 2001);
- Identifying the responsibilities that ensure all unit and/or team ground rules are compatible with achieving the stake, which provides stability and flexibility (Adler, cited by Choo and Bontis, 2002a);
- Enforcing the respect of the ground rules and arbitrate when inconsistencies and/or conflicts arise. The rela-

tionships between boss and subordinate (top-down), peer-to-peer, and client/supplier, must be defined into a contractual framework. (Koestler, 1964; Watzlawick, Beavin, and Jackson, 1967).

Contextual drivership protocols are the result of the integration between the situational leadership and the accountability approaches with the Principled Negotiation Approach developed at the Program on Negotiation of Harvard School of Law (Fisher, Ury and Patton, 1981, 1991; Lax and Sebenius, 1986).

Drivership makes reference to the constructs encompassing the noun drive, the verb to drive, and the suffix -ship (Collins Cobuild English Language Dictionary, 1987).

The noun drive evokes:

- The motivation and persistence to try new things and to do things well;
- A very strong need or desire in human beings that makes them act in a particular way;
- A special effort made by a group of people for a special purpose.

The verb to drive evokes:

- Operate and control (a device, a process) so that it works and goes where you want it to;
- Take somebody somewhere; supply power that makes a machine work.

The suffix -ship is added to nouns in order to form other nouns that represent:

- A person's position, status, office, skills, abilities;
- Relationship between people or things.

The individual, who engages in drivership, is called a driver. A colleague is the person with whom a driver initiates and regulates a contractual relationship, regardless of their hierarchical positions. Within the contextualist approach, drivership applies to the event in the context, not to a situation. The application of the contextualist perspective to situational leadership helped to design the contextual drivership protocols and to overcome the limitations of situationalism.

Situational leadership's theoretical foundations (Hersey and Blanchard, 1969, 1972, 1974; Hersey, 1984; Blanchard and Johnson, 1981) are linked to: Likert's (1961,1967) research on the effectiveness of different styles of management, Herzberg's (1959, 1966) research on satisfaction to work, Etzioni (1961) and French and Raven's (1959) research on the successful and effective use of power, and Lewin's (1950, 1951) use of force-field analysis. The Situational Leadership Model has been highly successful for several years in the USA as well as in Europe. It is based on two dimensions: task behaviors and relationship behaviors. To increase the effectiveness of their action, leaders are recommended to adapt their behaviors to the situation and to the features of the individuals they lead. In situational leadership, the most appropriate style of leadership is determined by making reference only to the readiness (combination of motivation and competence) of the follower in the given situation. Situational leadership has two main weaknesses:

- The stake of the leader is not taken into account;
- The reciprocal perception that the leader and the follower can have of the situation is not considered. The application of the contextual perspective to situational leadership helped to design the contextual drivership protocols and to overcome the limitations of situationalism.

A situation is a general set of circumstances and events manager and a worker discussing a quality issue). The same situation can take place in many different environments (a French Agri-food factory or a Chinese Agri-food factory, a French distributor of electrical supplies or a French distributor of textile products, an American assembly factory of Japanese cars or a Mexican assembly factory of European cars, etc.). Even if the situation can be described in the same way, if the environments are different, the contexts are different. A particular style of leadership can be effective in one context and ineffective in another context (French, 1959). Moreover, the effectiveness of leadership depends not only on the leader's style but as well on what the followers allow him to do (Crozier, 1963; Weber, 1978; Schroeder, 1992). Two factors are used to identify a specific context:

- What is at stake for the group of actors in the given environment (Fisher, Ury and Patton, 1981, 1991; Lax and Sebenius, 1986)
- What perceptions do the actors have of the events that take place in the situation (Pettigrew, 1990; Brouwers et al, 1997).

The protocols and tools developed to facilitate enacting the responsibilities of contextual drivership are discussed in part three. These protocols link driver's stake and colleague's reliability (combination of commitment and competence). The protocols provide a common framework useful for understanding differences in perception and agreeing on effective relationships to drive actions.

Operating Concepts of M3C: Reliability

In the contextual drivership framework, reliability is the assessed capability of an individual to persistently demonstrate commitment to the stake and to leverage the competencies appropriate to the activities required within the context.

The level of reliability is directly related to the commitment and the competencies of the colleague, as shown by the following formula:

Reliability = f (commitment of the individual, competencies of the individual)

The stability of reliability describes the likelihood for the colleague to persist in his logic of functioning for a reasonably long span of time.

The word reliability has been chosen to leverage the analogy with the reliability of a technological device (process, tool, or machine), as presented in Table 1.8:

Reliability of a technological device	Reliability of an individual
Device has the right features to carry out the requested activity	Competencies
Appropriate resources and maintenance are provided to keep device working	Commitment
Device can carry out the activity regularly over a consistent span of time, the duration of it depending on the conditions of the context	Stability
Device has a consistent interface with the operators and the other devices	Logic of functioning with others

Table 1.8—Rationale for the choice of the word reliability

The constructs of motivation/commitment and competencies/ability have generated discussion in the management literature in regard to the characteristics of an individual, which may be related to performance (Hersey and Blanchard, 1972, 1974, Tissier, 1988, Vroom, 1964; Hersey, 1984).

Vroom (1964) suggests looking at people's performance as the interactive result of ability and motivation, as depicted in the following formula:

Performance = f(Ability x Motivation)

From a systemic point of view, performance depends on the context and particularly on what the constraints are and on what is at stake. Therefore, an individual cannot fully control his performance, which is the main weakness of such an approach. What an individual does control is his capability to perform, his way of facing and carrying out a task, and the set of resources that he can leverage to have the best possible performance. Then the formula should be:

Capability to Perform = f (Ability x Motivation)

Hersey (1984) defines readiness as the integration of motivation and competence, and he uses that to assess the individual's present capacity to accomplish the assigned task. Readiness assessment is based on the observable behaviors and rated on a scale of four levels (R1 to R4) before deciding which type of leadership style has to be adopted. Table 1.9 shows the matrix of leadership style and task's control, used by Hersey to determine the individual's readiness.

		Providing Directive Behavior - Task Behavior			
		Low	Moderate		high
Providing Supportive Behavior - Relationship Behavior	High		Style 3 Participating Sharing ideas and facilitating followers in decision making	Style 2 Selling Explaining decisions and providing opportunities for clarification	
	Low	Style 4 Delegating Turning over responsibility for decisions and implementation			Style 1 Telling Providing specific instructions and closely supervise performance

Follower readiness			
high	Moderate		low
R4	R3	R2	R1
Able, willing, confident	Able, unwilling, insecure	Unable, willing, confident	Unable, unwilling, insecure
Follower directed task		Leader directed task	

Table 1.9—Hersey's Matrix of leadership style and task's control (Hersey, 1984)

Unfortunately, translating the matrix into action does not provide a consistent approach to the complexity of a work situation, as proved by the following example.

In level R3, the follower is assessed as able, unwilling, and insecure with the recommended leadership style as participating (share ideas and facilitate followers in decision making). Yet the follower, who has been designated as unwilling and insecure, directs the task.

In the M3C model, an effective driver would not allow an unwilling and insecure follower self-direct the task. The combination of unwillingness and insecurity generates a low level of commitment, which would be difficult to overcome merely by sharing ideas and facilitating the follower's decisions.

Despite its shortcomings, Hersey's model provides a good basis to develop the contextual drivership protocols presented in part three and the adoption of drivership practices.

1.2.6 On Accountability

Accountability involves the exchange of information between two parties where the one who is accountable and identified as agent explains or justifies his actions to the one to whom the account is owed, the principal (Swift, 2001; Gray et al, 1997; Woodward et al., 1996). The principal/agent framework is largely used in economic agency theory.

To protect his stake, the principal needs sufficient, accurate, understandable, and timely information to make decisions concerning the behavior of the agent (Swift, 2001).

The accountability approach assumes that agents cannot be trusted to act in the best interest of the principal when there is a divergence of stake between the two. In fact, agents might adopt opportunistic behaviors (self-interest seeking with guile) if they are not obliged to follow specific rules and are not controlled (Swift, 2001; Gray et al, 1997; Woodward et al., 1996; Hosmer, 1995; Huse and Eide, 1996; Hill, 1990, Williamson, 1975; 1985). Since trust cannot be assumed as a natural occurrence for the principal/agent relationship, the accountability approach recommends the establishment of a contractual obligation, formal agreements, and structural controls between the two parties (Swift, 2001).

Loyalty-focused management emphasizes building trust-based relationships as a means for establishing commitment and follow-through. Generating a "trust me" or "control you" culture establishes opaque relationships created by the power asymmetries of limited access to information. Each party has little choice but to believe whatever the other chooses to reveal about his actions, since commitment is rule-based rather than stake-based. When trust is deceived, it can increase the level of stress between the parties and provoke conflicts.

Management practices should move from a "trust me" or "control you" approach to a "tell me" and "show me" approach. The "tell me" and "show me" protocol, based on competence and integrity (Adler, cited by Choo and Bontis, 2002a), allows the principal and agent to protect common and individual Stakes by establishing the appropriate level of dynamic autonomy in their relationship.

Stake-based relationships prefer a "telling and sharing viewpoints" approach based on engagement, mutual experience, and dissemination of information (Swift, 2001). Accountability-based managerial practices should be designed into the contract and identify the appropriate positive and negative sanctions (Owen et al., 2000).

Accountability-based managerial practices reference two approaches in conceptualizing trust, the predictability of behavior, and goodwill (Swift, 2001).

The predictability of the agent behavior is based on past behaviors in specified contexts. Therefore, the principal has an expectation that in the future the agent will act as he has already done in similar contexts. Trust is predicated on the future behavior being that desired while distrust assumes the undesirable behaviors will continue.

This continuum of trust-distrust requires a shared past and the anticipation of a shared future. If one of these conditions is not satisfied, the principal must obtain a guarantee with acceptable consequences. At each point of the continuum, guarantees, standards, or legislation are necessary to establish an accountable relationship. Thus, trust becomes the enforcing of specific consequences associated with the acts specified in a contractual arrangement (Swift, 2001).

The reliance on an agent's goodwill to protect the stake references:

- The confidence in another party's reliability and integrity (Hosmer, 1995);
- The belief in another's credibility and benevolence;
- The belief in the reliability of another's premises. Goodwill as a substitute for trust is possible when the principal's and agent's stakes are mutually interdependent and stake attainment is contingent on cooperation. The recognition and acceptance of the interdependence and cooperation is a sufficient condition to ensure that the relationship works.

A split-trust continuum (Swift, 2001) incorporates the predictability of the agent's behavior and safety of the principal's stake into the operational use of trust.

The first continuum (Table 1.10), true accountability, presents a range of suspicion from distrust to lack of distrust. A lack of distrust means that the principal relies upon the predictability of the agent's behavior (Swift, 2001).

True Accountability (proxy for trust based upon duty to account by agent to principal)	
Range of Suspicion	
Distrust	Lack of Distrust
Assumption that the inconsistencies in the Agent's present behavior disallow for predicting the agreed behaviors.	Assumption that the consistencies in the Agent's present behaviors allow for predicting the agreed behaviors

Table 1.10—True accountability continuum (Swift, 2001)

The range of vulnerability (Table 1.11) is considered soft accountability because it is negotiated between the parties. The negotiation is based upon the parties' interdependence and the principal's perception of the agent's goodwill and the recognition the agreement cannot be enforced.

One anchor is unknown trust, which stems from principal's ignorance regarding the agent's past behaviors and intent. Known trust is based upon knowing the agent's reputation, opportunity for dialogue, experience, and personal engagement (Gray and al., 1997; Swift, 2001).

Soft Accountability (based upon interdependent relationships between principal and agent)	
Range of Vulnerability	
Unknown Trust	Known Trust
Ignorance of whether the Agent is trustworthy or not	Confidence in goodwill of the Agent's intent and behavior

Table 1.11—Soft Accountability continuum (Swift, 2001)

> True dialogue takes places only among equals. There is no dialogue across the boundary between masters and servants, for the master will listen only as long as his power remains intact, and the servant will limit his communication to utterances for which he cannot be punished. In fact, to recommend dialogue in a situation of inequality of power is a deceptive ideology of the powerful, who wish to persuade the powerless that harmony and mutual understanding are possible in society without any change in the status quo of power.
>
> (Baum, 1977: 43–44)

The relationship between the driver and the colleague can be analyzed using the principal/agent theory. The driver is the principal in charge of initiating actions to protect common and individual stakes. The colleague is the agent in charge of regulating or executing actions aimed to protect common and individual stakes.

There is no relationship between the hierarchical position and the role of driver or colleague. It may happen that a hierarchical superior has to act as a colleague to help a subordinate to take a driver position in a given action. In this case, the subordinate is the principal who initiates the action necessary to protect the stake at his level. The boss becomes the agent, who regulates the action and executes what is necessary for the principal to comply with the drivership duties and rights. Implementing the drivership approach in day-to-day operations requires recognizing the need of being mutually accountable in order to protect the common and individual Stakes.

Under the principal/agent framework, the two parties are bound by a contractual obligation. Should contracts be explicit, or can they remain implicit (Gray et al, 1997; Woodward et al., 1996; Swift, 2001)? The M3C requires adopted contracts between the driver and the colleague to be explicit, especially in terms of the relationship. Making the contract explicit does not mean that it needs to be formalized. Table 1.12 shows that formalization depends on the level of predictability and on the level of safety of the stake:

		Level of stake safety	
		Low (unsafe)	High (safe)
Level of predictability of other's actual behavior	High (predictable)	contract + guarantee + random control	contract or mutual agreement
	Low (not predictable)	contract + guarantee + continuous control	contract + random control

Table 1.12—Degrees of formalization of the autonomy contract in the drivership framework

The drivership paradigm uses the construct and level of autonomy to determine the contractual relationship binding a driver and a colleague. The level of autonomy depends on two factors: the reliability of the colleague in regulating or carrying out the activities initiated by the driver and the importance of the stake for the driver in a given context.

Autonomy is directly related to the reliability of the colleague, and inversely related to the stake of the driver, as shown by the following formula:

Autonomy = f (Reliability of the colleague, 1/Stake of the driver)

The perception the driver has of his own stake is biased by the perception he has of the reliability of the colleague and vice versa. If the driver grants to the colleague a high level of commitment and competencies, a reduction factor affects the perception the driver has of the degree and extent of control to exert with the colleague. If the driver assigns a low commitment and low competencies, the perceived need to control increases. For the intermediate situations (low commitment high competencies or high commitment low competencies) variations occur in the nature of the driver's behavior.

The inconsistency of Hersey and Blanchard's model can be overcome by using the autonomy curve, explained in detail in part three. The autonomy curve is the integration of the contributions of Hersey and Blanchard (1969, 1972), Hersey (1984) and Tissier (1988) on the situational leadership approach, the principled negotiation approach developed at the Program on Negotiation of Harvard School of Law (Fisher, Ury and Patton, 1981, 1991; Lax and Sebenius, 1986), and the development of the concept of accountability.

The autonomy curve method ensures the necessary systemic links between the needs of the driver and the resources of the colleague. In addition, it can be used to assess the overall functioning of a team in carrying out a task, an action, or a project.

Autonomy is not the same as independence:

- Autonomy references "the ability of an individual to make own decisions about what to do rather than being influenced by someone else or told what to do" (Collins Cobuild English Language Dictionary, 1987). Therefore, in a given context, the lower the Autonomy, the higher the importance and the frequency of others' decisions, and vice versa;

- Independence references "the fact of existing and acting separately from other people" (Collins Cobuild English Language Dictionary, 1987).

Accountability and autonomy are the result of a learning process and the enforcement of ground rules. Drivers have to play the role of facilitators and advisors to foster the autonomy of the colleagues through the development of their competencies and by providing meaning to their actions (Noguera, 2003).

1.2.7 On Commitment and Motivation

The main goals of M3C are to create the conditions to support individuals' commitment and enhance group cohesiveness, since they are key factors in controlling the behaviors of its members.

The form and structure of a group depends, in part, on the nature of the internal/external communication of its members and the lines of influence, power, and authority existing among them and with other external players (Lewin, 1950, 1951). For these reasons, several historical contributions have been taken into account in designing the model.

Motivation

Motivation is a sensitive issue in implementing and sustaining change. Motivation is a process governing choices made by persons among alternative forms of voluntary activity (Vroom, 1964). Motivation is concerned with three fundamental questions related to behavior regulation: what energizes action, how action is directed, and to what extent action is under voluntary control (Deci, 1992).

Motivation to work is the willingness of an individual to exert high level of effort to reach organizational goals, conditioned by the effort's ability to satisfy some individual needs.

Motivation is an intrinsic energy that is activated when an individual finds out in his working environment the appropriate reasons to invest himself into those activities that fit to his expectations and desires (Tissier, 1988).

The Hawthorne Studies demonstrated the unexpected consequences of an action instituted by management. Production output was influenced more by the perceived relationship of the individual with the supervisor than by the type of working environment (Mayo, 1949). The workers responded positively regardless of the working condition change. The workers perceived each change as a concern for their welfare, even if it wasn't intended as such.

Motivation increases if a group believes it has been allowed to choose its own goal, within the context of the stake. This belief facilitates the initiation and stabilization of the change process. Employees who are given the opportunity to play a role in goal-setting and decision-making affecting their work will accept change more readily than those who are catapulted into change without any opportunity for the exercise of free choice (Coch and French, 1948). Motivation is considered an inner state rather than the result of a person's actions to make another individual behave as the former wishes. Concerning motivation to work, it is possible to identify two main theoretical approaches:

- Content theories focus on what motivates people at work. Maslow (1943, 1968) and Herzberg (1959, 1966) focused on the individual's need satisfaction;

- Process theorics focus on how to motivate people at work. Vroom (1964) and others focus on the valence, the expectancy, and the force that the individuals perceive in the outcome of their work.

Commitment

Commitment to work has historically not been considered a major topic in managerial training. Buchanan (1974, p.533) defines commitment as "a partisan, affective attachment to the goals and values of an organization, to one's role in relation to goals and values, and to the organization for its own sake, apart from its purely instrumental worth. Methodologically, commitment consists of three components … (a) identification, absorption as one's own the goals and values of the organization; (b) involvement, psychological immersion or absorption in the activities of one's work role; (c) loyalty, a feeling of affection for and attachment to the organization."

Salancik (1977b, p.62,64) states, "Commitment is a state of being in which an individual becomes bound by his actions and through these actions to beliefs that sustain the activities and his own involvement … Three characteristics bind an individual to his acts and hence commit him. They are the visibility, the irrevocability, and the volitionality of the behavior."

O'Reilly and Caldwell (1985, p. 467; 1980) demonstrated empirical support for the attribution nature of commitment in a field setting. Their findings support commitment resulting from irrevocable and volitional acts the individual engaged in during the job choice process.

According to Porter, Steers, Mowday, and Boulian (1974), commitment consists of (a) a belief in and acceptance of organizational goals and values, (b) the willingness to exert effort towards organizational goal accomplishment, and (c) strong desire to maintain organizational membership.

Singh and Vinnicombe (2000) found role and the sex differences between male and female engineers and the engineers' managers impacted the meaning of commitment. Eight major classes of meaning were identified: (a) task delivery, (b) put yourself out, do extra, (c) involvement, (d) quality, (e) be proactive, use initiative, (f) ready for challenge, (g) creativity, innovation, (h) add value/business, customer awareness.

The individual's social interaction with peers or superiors in the workplace is a necessary condition for the establishment of commitment while job achievement and hierarchical advancement are factors likely to enhance commitment (Buchanan, 1974, p.534).

As Senge (1994, p.494) points out,

> If the members of your organization are committed, the leadership ... comes from each member. People continually learn and improve their own and others' performance. The management's task is to manage ideas, coordinate resources, and create quality work environment—not to generate motivation ... Commitment builds when people are an active part of the experience of creating something they value together. Using common language, symbols, and metaphors, which evoke positive emotions also help bring people together ... To build community in an organization under the stress of unrelenting change, serious attention must be paid to mutual commitment ... Many employees are willing to commit themselves to a truly engaging purpose, larger than just personal self-interest ... The employees also seek concrete evidence that the collective enterprise is committed to them. They prefer to see tangible results from the investment of their efforts.

When extrinsic factors weaken or disappear and hygiene factors deteriorate, committed individuals continue to leverage their best resources to provide effective contribution to the team's actions.

> Once an individual is personally committed, the efforts to get the target creates an internal tension that contributes to the achievement. If the individual meets some obstacles in her way to accomplish the task, most likely he will not say 'I am stuck,' rather he will try harder to eliminate the obstacles ...

> Committed individuals will not give up when they will face difficulties.
>
> (Zeigarnik, 1927)

The persistence to achieving the stake is a critical factor when change, especially change requiring institutional orientation, is instituted.

Because committed individuals are aware and proud of their contribution to the community and want to abide by the "moral contract" of "community," they are capable of facing the tribulations of the existing context for longer periods of time. Committed individuals prioritize their actions to protect the common stakes instead of seeking the satisfaction of their own needs or interests (Cialdini, 2001).

Individuals alter their beliefs, often mindlessly, to make them consistent with their acts (Cialdini, 2001; Piattelli Palmarini, 1995). To the extent the change process is nothing but the implementation of new ideas starting from acts, these ideas will be rationalized and verbalized as acceptance of the change (Watzlawick, Weakland and Fish, 1974).

Change by commitment has a success rate higher than change by external motivation, because commitment requires explicitly stating the actions and accomplishments to which the individuals must bind them (Joule and Beauvois, 1987, 1998).

Explicit commitment helps individuals clarify their relationships and roles in the team to consistently carry out the actions required by the change process.

For the driver, leveraging commitment requires facilitating the colleague:

- Verbalize his feeling and understanding of the context;

- Recognize own, others,' and common stakes;
- Describe the his satisfiers and dissatisfiers;
- Explicitly state his personal choice regarding satisfiers and dissatisfiers;
- Identify existing competencies and those that must be developed to comply with his commitment;
- Act (demonstrate behaviors) appropriate to the verbalized commitment;
- Relate every action to accomplishments.

Commitment makes reference to the employee's decision to participate by accepting the field of meaning of the company (Spender, 2004). Therefore, since it makes explicit the employee's sense of identity with the enterprise and its mission (Nonaka, 1991), it is a critical factor in determining an individual's reliability.

The M3C model synthesizes the concepts of group cohesiveness, motivation, satisfaction, and commitment to support a holistic, transformational process, with the goal of developing a community of committed, competent individuals dedicated to achieving the stake.

Commitment is the individual's state of being, which is grounded on his awareness of the context. No matter what the satisfiers and dissatisfiers factors present in the context, it leads the individual to explicitly and voluntarily leverage his own physical, intellectual, emotional, and intuitive resources to protect his own and his team's stakes, and to contribute to his own and his team's actions and accomplishments (Herzberg, 1958, 1966; Becker, 1960; Buchanan, 1974; Porter, Steers, Mowday and Boulian, 1974; Salancik, 1977a, 1977b; Wiener, 1982; Meyer and Allen, 1997; Kanter, 1968).

1.2.8 On Competencies Development

The effective implementation of the M3C is predicated on the development of an individual's competencies. New or enhanced competencies encourage individuals to experiment with new approaches, to learn from others' experience and best practices, and to transfer knowledge quickly and efficiently throughout the organization. A key feature of a learning organization is the implementing of change in a cooperative manner (Garvin, 1993).

Useful tools to implement competencies' development strategies in support of M3C job are:

- Enlargement and its derivative job rotation and job enrichment (Herzberg, 1966);
- Portfolio activities approach (Handy, 1989);
- Employability approach (Aubrey, 1994);
- Knowledge creation steps (Nonaka, 1991; Nonaka and Takeuchi, 1995).

In French business literature, competencies are the combination of savoir, savoir-faire, and savoir-être. Savoir is related to the theoretical knowledge acquired by studying, savoir-faire to the know-how acquired through reflective experience, savoir-être to behaviors (Aubret, 1993; Le Boterf, 1994; Minet, Parlier and de Witte, 1994; Grimand, 1995; Klarsfeld and de Saint-Giniez, 1999; Biscaccianti and Feron, 1999).

Table 1.13 shows the four different types of erudition referenced by ancient Greeks (Baumard, 1996):

	Epistémé (Knowledge)	*Techné* (Know-how)	*Phronesis* (Behaviors)	*Mètis* (Competence)
Definition	Theoretical Generalization	Ability for Task Accomplishment	Social Wisdom	Conjectural Intelligence (capability of dealing with ambiguity)
Goal	Truth	Method	Wisdom	Result
Approach	Abstraction	Observation and Action	Socialization	Combination
Field	Universal	Systems	People	Context

Table 1.13—Different types of erudition in Ancient Greece (Baumard, 1996)

Erudition is considered an on-going process to achieve Métis, the final step where all three previous dimensions are combined into a holistic complex.

Quinn, Anderson, and Filkenstein (1996) recommend the fostering of four types of knowledge to ensure the development of high potential people:

- Cognitive knowledge (know-what, corresponding to Epistémé);
- Advanced skills (know-how, corresponding to Techné);

- Self-motivated creativity (care-why);
- Systems understanding (know-why, corresponding to Métis).

The four steps of the knowledge creation cycle (Nonaka, 1991; Nonaka and Takeuchi, 1995) can be analyzed using the four types of knowledge identified by the ancient Greeks:

1. Learn the tacit knowledge (socialization phase, where the Phronesis and Techné dimensions play a major role);
2. Translate tacit knowledge into explicit knowledge (articulation phase, where Techné and Epistémé are leveraged);
3. Standardize this knowledge into protocols (combination phase, devoted to the Epistemé consolidation);
4. Enrich own tacit knowledge base (internalization phase, where Métis is properly achieved).

Individuals can best protect their employability when they are able to periodically change the mix of their activity portfolio (Aubrey, 1995; Handy, 1989). The appropriate use of the existing activities, the development of new ones, and the dropping of activities that are no longer useful to improve competencies best keep competence relevance up to date.

The analogy of the product life cycle is useful to explain the benefit of such an approach:

- At an early stage, competencies must be discovered, acquired, and explored through a controlled investment (development activities equivalent to the product development phase);

- As such competencies become distinctive and provide increasing added value, they must be intensively leveraged (key activities equivalent to the product growth phase);
- When competencies are sufficiently mastered and the potential for improvement and increase in performance has decreased to a stable, low level, they must be used to minimize the use of resources to produce stable added value (routine activities equivalent to product maturity phase);
- In their final stage, obsolete or low, added-value competencies must be abandoned to free resources for the development of new edge competencies (sleeping base activities equivalent to product decline phase).

Operating Concepts of M3C: Competence

Competence is the balanced and appropriate combination of knowledge, reflective experience (know how), and behaviors that an individual uses to carry out a given activity in a specific context (Nonaka and Takeuchi, 1995; Senge et al., 1994).

To effectively deal with the complexity of changing contexts, individuals must be able to use all available, tangible, and intangible resources to modify their existing competencies and to create new ones. Competent individuals have the capability of intelligently understand phenomena (even when ambiguity is high) and use all knowledge, skills, and behaviors to successfully solve problems and achieve results. Individuals who have developed Mètis are able to act appropriately, even if they do not have all the cognitive or experiential elements necessary to make a decision.

1.2.9 On Creative and Innovative Thinking

Innovation is the implementation of new combinations of different resources in the firm (Schumpeter, 1967) and requires breaking free of the usual relations between the elements of a system (Alter, 1993). Technological innovation and social innovation both contribute to introduce change into processes and products.

What matters is not only the newness of the processes and the products, but as well how these creative outcomes are accomplished, how they are promoted, how they shape the perception, and the practices of the other individuals involved in or impacted by the change (Ford, 2000).

Originality, freshness of perception, and divergent thinking are desirable personality traits. However, from a managerial perspective, they foster innovation and creativity only if they have some form of public recognition leading to a modification of the social outcomes (Vissers and Dankbaar, 2002; Csikszentmihalyi, 1991). Social innovation consists then in the codification of a specific type of interaction to shape new and different organizational processes (Harrison and Laberge, 2002) while creativity is linked to personal attributes (Vissers and Dankbaar, 2002).

Innovation triggers change and provokes resistance and conflicts because it requires more than the performance of routine tasks, which may disrupt the identities of the actors (Harrison and Laberge, 2002).

Caring about individuals' and teams' working conditions is then necessary to foster innovation in products and processes through individuals' and teams' creativity (Hertzberg, 1959, 1966; Vissers and Dankbaar, 2002). The appropriate level of formalization of the organization must be found to preserve the autonomy of the individuals and the teams in charge of driving the change. Highly formalized organizational processes better suit incremental innovation, and more organic structures enable teams to produce radically new ideas, concepts, processes, and products (Woodman et al., 1993).

March (1991) distinguishes between knowledge exploitation and knowledge exploration. The former is the primary task of the formal structure, requires high levels of specialization, the maintenance of standardized routines, and an emphasis on efficiency and reliability. The latter not only requires less specialization, an emphasis on problem-solving rather than routine, and low levels of formality but also coordination and knowledge-sharing among different individuals and across different functions and departments than is required for exploitation.

Innovation in processes and products is then more easily established through the construction of consent rather then through coercion (Tuckman, 1995); creativity is enhanced by group diversity and multidisciplinary cooperation (Jehn et al., 1999; Knights and Wilmott, 1997; Jassawalla, Sashittal and Hemant, 1998). Diversity ensures complementarity (Jackson, 1996) while multidisciplinary cooperation allows to bridge gaps resulting from disciplinary specialization (Nissani, 1997). The more multidisciplinary a team is, the more it will develop absorptive capacity; that is, the capability of re-creating knowledge on the basis of this prior knowledge (Cohen and Levinthal, 1990; Liyanage and Barnard, 2002; Spender, 2004). A diverse multidisciplinary team will be able to enhance the prospects of co development activities and draw upon different fields of competence (Vissers and Dankbaar, 2002).

To enhance cross-functional skills integration, facilitate information sharing, and prevent downstream issues, the driver of the innovation and change process has to give special attention to how multidisciplinarity is managed (Brown and Eisenhardt, 1995; Jassawalla and Sashittal, 1998). Research shows that to meet the novelty and complexity in innovation and change processes, senior management has to teamwork to build and spread a shared common vision; middle management has to ensure cross-functional interaction to meet strategic challenges; supervision has to leverage organizational competencies (Isaken and Lauer, 2002).

Facilitation is then necessary to help the team to explore new perspectives while keeping focused on the global performance of the process (Vissers and Dankbaar, 2002; De Dreu and Beersma, 2001) and to overcome frequent diversity and multidisciplinary pitfalls like communication, coordination, and efficiency problems (Reagans and Zuckerman, 2001).

Building on Witte's (1973) original model, Hauschildt and Kirchman (2001) research substantiated then the interest of having three types of promoters to cooperate and drive an innovation and change process:

- The "technology promoter," whose role is to provide specific technical knowledge to educate people and remove the barriers due to ignorance and uncertainty;
- The "power promoter," whose role is to leverage hierarchical position and overcome people's unwillingness to involve themselves in the process;
- The "process promoter," whose role is to use relational influence to mediate between people and overcome the barriers of non-responsibility and indifference due to organizational and administrative inertia.

Very important is the action of the process promoter, who

> Is always needed when particularly complex material is to be handled [because he is] able to translate the language of innovative technology into the language traditionally spoken and understood in the firm. He is the champion of innovation, able to turn an idea into a plan of action. He has diplomatic skills and

> knows how to approach and win over different types of people on a one-to-one basis.
>
> —(Hauschildt and Kirchman, 2001, p.42)

Innovation and creativity can be fostered by group interaction. Jassawalla, Sashittal, and Hemant (1998, p.50) define interaction as "the process by which participants in product-innovation settings (a) articulate and make explicit to others their deeply held, tacit knowledge about their areas of expertise and discuss their core values and aspirations; (b) intensely exchange explicit information in ways that expand their way of thinking, informed by the diverse talents, knowledge bases, and aspirations that exist in their settings; and (c) integrate the new, emerging, explicit knowledge into their tacit understanding of the new-product task." The tacit-explicit interactions largely determine the extent to which participants make use of others' knowledge, build on others' creative insights, and harness the synergistic potential of product-innovation settings.

Highly innovation-supportive cultures foster interaction by encouraging a culture of inclusion to overcome "us versus them" thinking. Frequent formal meetings help to share information, exchange and develop ideas, express disagreement, manage conflict, ensure that diverse voices are heard and discussed in open forums, and that a shared understanding emerges among participants (Jassawalla, Sashittal, and Hemant, 1998). From this perspective, anonymity encourages people to express dissenting opinions, which stimulate sense-making attempts (Sosik et al., 1998).

De Bono's (1985) approach to parallel thinking can be useful to create the conditions for innovation through problem-finding and prediction/anticipation (Choo and Bontis, 2002a, p.6). His starting point is that dialectic debate on ideas is the biggest obstacle to creativity and innovation because it promotes confusion in mental fields. He defines a mental field as a thinking role to be played as a specific perspective that can be taken when dealing with an issue. Table 1.14 shows how each mental field, associated with a color, describes attitudes and behaviors coupled with his complementary mental fields:

De Bono's six thinking hats do not cover all the mental fields, but they represent the main ones, allowing then people to distribute interpretations, canalize the attention, leverage symbolism and metaphors, and set up ground rules for dialogue and exchange.

The first aim of this approach is to simplify thinking by handling one thing at the time, instead of mixing emotions, rationality, creativity etc. Its second aim is to create the condition permitting the shift from one mental field to another by creating a symbolic code for attitudes and behaviors.

White hat *Neutrality and objectivity* Reference to data and facts	**Red hat** *Bias and subjectivity* Reference to emotions, feelings, irrational aspects, intuition
Yellow hat *Positive view* Reference to opportunities, advantages, effective constructive solutions	**Black hat** *Negative view* Reference to threats and drawbacks
Green hat *Creative approach* Reference to new ideas, divergence, exploration out of the box	**Blue hat** *Rational approach* Reference to process's consistence, reflective thinking, logical criteria, convergence, implementation

Table 1.14—de Bono's Six Hats technique mental fields (de Bono, 1985)

PART TWO: THE M3C—MODEL OF COOPERATIVE CONTEXTUAL CHANGE

Historically, the change process has been framed using constructs, which constrain its appropriateness and sustainability.

Therefore, to improve the probability of successful implementation and sustained change initiative, the constructs framing change need to be altered. As proposed by communicative therapy and contextualist approaches, effective change requires a shift in the existing paradigm, which supports illogical interpretations of the context. Reframing the perception people have of a context changes their behaviors and the actions deemed necessary for achieving the goal. (Watzalawick, Weakland and Fisch, 1974; Weick, 1979; Giddens, 1979; Pettigrew, 1987; Brouwers et al., 1997).

Table 2.1 presents questions regarding change and the answers generated from the traditional paradigms and the alternative M3C paradigms, which will be discussed later.

Question	Traditional paradigm	M3C paradigm
What drives change?	Reaction to avoid failure	Pro-action to protect the stake
How are ground rules established to create the framework for the process of change?	Authority and expertise	Co-building and co-opting
What relationships drive the interfaces among the parts of the organization?	Leadership responsibilities and management responsibilities	Initiating responsibilities and regulating responsibilities
What is the mindset for action?	Problem-solving	Solution-orientation

Table 2.1—Traditional and M3C constructs to change

The traditional change paradigm accepts instability as a necessary pre-condition (Chandler, 1973, 1976, 1989) regardless of the theory (E or O). Instability as a consequence of failing to achieve a desired state appears to be significantly different from the instability experienced, because the system/individual desires to improve the existing "successful state" (externally forced or internally mandated). Instability increases the stress experienced by the organization and the individuals within it; as instability increases, so does the probability of resistance, unless individuals fully understand the value of the change initiative and accept it as their own, which is the consequence of the decision to participate (Spender, 2004).

Stability and/or instability in implementing change must then be considered from an alternative perspective, which incorporates constructs of co-building, co-opting, and cooperation in implementing and sustaining change initiatives. The alternative perspective proposed by the M3C is that the change process can be fruitful only if the organization and the individuals acknowledge and accept a common point of stability from which to initiate change.

2.1 The Four Elements of Dynamic Stability

The model of cooperative contextual change posits that dynamic stability must exist to create the conditions favorable for successful change (Senge, 1994 et al.; Desikachar, 1999).

Dynamic stability is both a process and a state of being that allows the organization to modify its functioning while retaining and enlarging its identity. As a process, dynamic stability encourages alterations in behavior and relationships, which facilitate achieving the stake. As a state of being, dynamic stability provides the security of identity, even as the organization and the individual evolve. M3C recognizes that for the individual, experiencing a sense of community is essential for commitment to the change initiative.

The organization as a whole as well as the individual must be presented with intellectual and emotional anchors that provide a sense of security as the tide of change rises. The critical intellectual anchor is the understanding of how the organization and the individual will function within the future context. As the individual's awareness of his functioning within the new organizational context increases, acceptance of the change initiative increases (Anderson and Ackerman Anderson, 2001). In addition, if the individual understands the rationale for the change and his role, the probability of its success increases. Acceptance requires an explicit commitment to involve oneself in the process of change, which includes accepting responsibility for personal and team actions.

As mentioned above, dynamic stability is a both a state of being and a process, similar to the status of atoms. Physicists discovered that the atom is only momentarily isolated in time and space before changing into a process. Individuals typically see change as traumatic, transformational, and requiring a significant alteration in their existing condition. In point of fact, most change is incremental (e.g., weight gain, aging, and knowledge and skill acquisition). Because individuals are unaware of incremental gain or loss, they don't resist it.

If individuals understand that change may be both dynamic and stable, they are more likely to accept that they can control what happens to them and how they respond within the context. This acceptance requires an increase in the level of awareness of their function, the function of the context within which they exist, and of the environment within which the context exists.

For dynamic stability to exist, individuals must be responsive to the existing context and proactive to the anticipated and/or desired context. A responsive orientation effectively and efficiently integrates the incremental change required to resolve existing issues. A proactive orientation prepares for future changes that have an increased probability of occurring. Being proactive reduces uncertainty and institutes behaviors that are more likely to be successful as the future context evolves (Denison and Mishra, 1995; Collins, 2001).

The M3C model posits that dynamic stability is achieved through a sense of community within which the individual must be involved. When people experience the process of community, this enables them to understand and use the four stable elements of dynamic stability: the stake, the ground rules, the systemic relationships, and the solution-oriented mindset.

2.1.1 The First Element of Stability: Explicit Stake

The stake is the most valuable tangible or intangible asset that an individual or a community wants to protect with their action. The individual, the organization, and the community have an intangible or tangible asset, which is considered invaluable and essential and must be protected (i.e., quality of customer service). Understanding the criticality of this asset—the stake—to the organization and the individual is essential in any change initiative. Therefore, the explicitness of the stake can be used to sensitize individuals to their context and the potential consequences of the change initiative. Heightened awareness encourages the identification of the parts of the environment that must be integrated into the change initiative (Watzlawick, Weakland, and Fisch, 1974).

The impact of an actual or forecasted significant event on the relationship between the community's context and its environment should be assessed in reference to two questions: "What is at stake for the community?" and "Is the stake in danger?"

Will the existing or future community's context match the existing or future environment? If the community's context is or will be out of alignment with regard to the environment, the significance of the misalignment must be determined. If the community's stake is in danger, the community must begin a systematic process that will culminate in a change initiative that protects the stake.

Recognizing that the external environment is dynamic, the community will need to develop a plan that allows and even encourages deviations in response to or anticipation of the changing conditions of the environment and the community's context. The appropriateness of the selected solution and its associated process activities must be built within the community. An internally developed change initiative permits the control of deviations through the establishment of ground rules, which frame the activities, responsibilities, and relationships of all individuals within the community.

2.1.2 The Second Element of Stability: Co-Built and Co-opted Ground Rules

Ground rules define the way the actions are carried out within the context of the community, as well as the rules of engagement between the community and the external environment. Ground rules clarify the interfaces between the elements of the systems, the overall philosophy of their interaction, and the different roles and responsibilities. They identify and manage the drifting of functioning, the way trade-offs are made to satisfy divergent interests, the decision-making process, and the way actions will be taken to protect the cohesiveness of the community (Grant, 2001).

To ensure the progress of change, sustainable cooperation must exist within the community. Therefore, an essential ground rule is that, in interaction, individuals will treat each other with respect as persons while acknowledging that disagreement may exist in the activities linked to their functional roles (Rosenberg, 1999; Grant, 2001; Choo and Bontis, 2002a).

The prime ground rules are:

- First, ground rules must be developed through a two-step process of co-building and co-opting;
- Second, all ground rules must be constantly measured against the stake and the environment. If changes in the environment will affect the stake, the context changes, and the first prime ground rule must be invoked to protect the stake (Koestler, 1964).

A corollary of the prime ground rules is that management by exception cannot be permitted because it cannot be regulated.

Any ground rule must be explicitly associated to a sanction, positive or negative. The sanction must be co-built and co-opted at the same time as the ground rules.

2.1.3 The Third Element of Stability: Systemic Relationships to Initiate and Regulate Actions

Organizational alignment is necessary to protect the stake: the goals, values, beliefs, structure, and processes are all focused on achieving the common stake.

Alignment requires a systemic relationship among the members of the community: each individual must accept and take responsibility for initiating and regulating every action necessary to achieve and protect the stake (Senge et al., 1994). Any individual recognizing a deviation from or an innovative improvement process in the protection of the stake, must present his recommendations to the community. Thus, initiating responsibility (IR) adds value to the action of each actor, and granting value to his action enhances the meaning of what he does. Meaning is the intrinsic value that pushes each part to increase the global performance (efficiency + effectiveness) of the system as a whole (Anderson and Ackerman Anderson, 2001).

Regulating responsibility (RR) requires any individual to make sure that the activities necessary to protect the stake are performed consistently with the agreed decision of the community. When each actor is assigned and accepts regulating responsibilities, relevance is added to his actions. Relevance speaks to the contribution of each actor to the viability of the other levels of the system, the way each part fits into the larger picture and serves the larger systems (Anderson and Ackerman Anderson, 2001).

When an individual decides to enact initiating and regulating responsibilities, he must ensure the second prime ground rule is met.

2.1.4 The Fourth Element of Stability: Solution-Oriented Mindset

In any system, when new learning is put into place, initially there will be a degradation of performance efficiency and effectiveness. As individuals acquire new knowledge and skills, errors and mistakes in omission and commission will occur. Acceptance of this reality requires the institutionalizing of each individual's right to make errors as well as the duty of learning from them. However, repetitive errors must be considered a failure to accept initiating and regulating responsibilities, which invokes ground rule sanctions, or a lack of competence, which leads to re-training. The granting of responsibilities and learning experience seek to establish a solution-oriented mindset.

A solution-oriented mindset requires four conditions:

- Individuals capable of presenting a solution for the issue rather then offering an explanation as to why the issue exists (Watzlawick, 1983);
- The offered solution must go through the process of co-building and co-opting (Schutz, 1958, 1966, 1982, 1984);

- The solution must meet the requirement of protecting the stake (Fisher, Ury and Patton, 1981, 1991);
- The solution must be ecologically optimal, in that it promotes the overall well-being of the community without reducing the well-being of any of the parts (de Saint Paul and Tenenbaum, 2002; Fisher, Ury and Patton, 1981, 1991).

These four conditions are necessary for fostering cooperation and innovation among the community's interdependent roles.

2.2 THE M3C—MODEL OF COOPERATIVE CONTEXTUAL CHANGE

The model of cooperative contextual change (M3C) describes the understanding reached after years of management, consulting, and teaching experiences, which involved helping organizations build dynamic stability and successfully engage a change process.

The M3C is not a theory. A theory attempts to explain or interpret why things happen as they do, based on a hypothesis, which must be proven in order to be held true. A model describes how a system or a process works based on a pattern of already existing events, behaviors, and/or ways of acting. The significance of a model is dependent upon its capacity to be translated into a learning process and outcome. (Hersey, 1984; Collins Cobuild English Language Dictionary, 1987; Watzlawick, Beavin, and Jackson, 1967).

A model promotes a precise language that people can both understand and use to interpret what they are experiencing in their relationship. Therefore, a model can be used to meta-communicate the implementation of a type two change (Watzlawick, Weakland and Fisch, 1974; Tissier, 1988).

Change process models are tools that provide guidance during conscious process thinking. They are both action oriented and result producing. Models organize the change process activities into sequential, developmental units to facilitate time allocation per stage and process feedback, which increases the probability of achieving the desired outcomes. A process model drives change by altering the system's dynamics and by providing suggested plans of action for how to change an organization over time (Anderson and Ackerman Anderson, 2001).

The M3C links several change process models. Its scope is sufficiently broad to allow its application to a wide range of situations. The M3C establishes conditions, which encourage co-operation, commitment, and the leveraging of resources that protect the stake of the organization. Because of these conditions, the M3C has face validity that encourages acceptance and implementation by any position in the company. The model is applicable to any organization (governmental, non-governmental, for profit, not-for-profit), regardless of the industry, domain, or size, for implementing change at any level (individual, team, department, company). This book presents the conditions, processes, and procedures of the M3C in regard to implementation within an organization.

The model is designed to ensure that an organization engaged in the process of change achieves its goals. Need for change is not a required condition to apply M3C, but positive change is a consequence of its implementation. In fact, organizations that are experiencing success are encouraged to implement the model, because it can further enhance success. The M3C paradigm is, "Failure is no longer a required condition for investing in change."

The model consists of four main modules, as shown by Table 2.2: Contextual Assessment (CA), Community Norming (CN), Contextual Drivership (CD) and Cooperative Innovation (CI). A full explanation of the protocols and tools useful to implement each module will be presented in chapter three.

Module	Context's Assessment (CA)	Community Norming (CN)	Contextual Drivership (CD)	Cooperative Innovation (CI)
Stake	Checking the applicability of the model	Setting the team and ensuring its cohesiveness	Initiating and regulating actions	Producing meaningful solutions
Purpose	*Install the first element of stability: explicit stake*	*Install the second element of stability: co-built and co-opted ground rules*	*Install the third element of stability: systemic relationships*	*Install the fourth element of stability: solution-oriented mindset*
	Bring awareness of the need for change	Create membership to enhance team's stability	Ensure individuals' powerfulness to enhance team's performance	Stimulate creativity and innovation to enhance team's competitiveness
Goals	To assess the context of the organization and to identify the Stake of the change process To align organizational goals at any level To detect and reduce inter-organizational and interpersonal frictions, resistance and blockages To co-design with the Driver the process to implement the model and the tools for measuring progress	To set-up ground rules to facilitate cooperative teamwork To put together divergent interests to protect the identified common Stake To establish a consensual decision-making process to avoid implementation failures To make collective and individual commitments explicit to ensure long-term support to action plans	To fine-tune system's functioning, interfaces' quality and activities' consistency to organize the complexity of interdependent actions To increase people's Reliability and establish autonomy-based relationships among the actors To facilitate feed-back at any level to integrate the constraints of the context into team actions	To create and to support the reflex of positive contribution by focusing on solutions instead of problems To concentrate resources to tackle issues from different points of view in a cooperative way To encourage forward thinking, knowledge sharing and best practices seeding To implement innovative work team technologies and processes

Table 2.2—The four modules of the M3C

The implementation of the M3C is based on a cooperative, participatory learning process. As the individuals learn, they are asked to change their constructs and their behaviors. Discrepancies may happen between the intellectual understanding of the individuals and the actions they carry out. If the individual is capable of demonstrating behaviors that are supportive of the new constructs, acknowledgment of this capability can constitute an element of stability in the change process. Therefore, the driver should periodically have individuals demonstrate their awareness of the goals and the actions they are taking to achieve them (reference the goals identified in Table 2.2).

2.3 THE LEARNING PROCESS TO IMPLEMENT M3C

Learning to learn is a necessity in all types of change but especially transformational. Individuals must unlearn old practices and learn new protocols (Senge et al., 1994). Three different approaches are available to facilitate the "learning to learn" process: techniques based, operating procedures based, and theoretical knowledge based (Blake and Mouton, 1985).

The techniques-based approaches focus on managers learning and using new management tools such as MBO, job enrichment, TQM, BPR, kaizen, kanban, ISO standards, etc. These approaches may succeed in provoking first-order change, but they often fail in creating second-order change (Watzlawick, Weakland and Fisch, 1974). Even if "change" tools are implemented, people may neither question nor reconsider their own management style or understand change mechanisms (Korzybski and Kendig, 1942).

The operating procedures-based approaches are centered on team-building, team dynamics, and teamwork facilitation driven by an external facilitator. If the tools are merely learned in artificial contexts (simulations, role plays, or games) but not internalized by the participants, they will not be implemented in day-to-day operations (Blake and Mouton, 1985). Operational approaches may provoke first-order and second-order change, but they do not ensure persistence in time. After the facilitator leaves and people return to normal working conditions, lack of internalization and ownership inhibits commitment.

After the training, the presence of an "enforcer of protocols" within the organization is a key factor for success. The preference is for the individual to be the external facilitator, the driver of the change, or any other person duly appointed as the "protocol enforcer." The "protocol enforcer" is entitled to ensure the enforcement of the ground rules and to drive the individuals and the team in the implementation of the new protocols.

The theoretical knowledge-based approaches recommend that people first learn a theoretical basis for performing the behavior. Then they put the new behaviors into practice (Blake and Mouton, 1985; Hersey and Blanchard, 1969, 1972, 1974).

Such approaches may provoke both first-order and second-order change, but they often fail because of inconsistencies between intellectual understanding of the theory and actual implementation of the required behaviors (Watzlawick, Weakland, and Fisch, 1974).

The learning process designed to implement the M3C integrates these three approaches. Its purpose is to create the most favorable conditions for a group (two to several hundreds) to work effectively together. Change becomes the consequence of its implementation.

The successful and effective implementation of the M3C is largely dependent on the embedment of the deutero-learning into day-to-day operations. As a general recommendation, valid for all the four modules of the M3C, the first two months following the introduction of learning to learn is critical for sustaining the change initiative. The driver should encourage people to operationalize the learning by allocating specific time and resources to practice, by temporarily reducing performance and productivity requirements and by granting tolerance toward mistakes.

The institutionalization of a continuous follow-up process is necessary to make sure that the implementation of the acquired protocols is carried out in the appropriate way. When mistakes, errors, or deviations occur or can be anticipated, corrective actions can be rapidly taken and adaptations put in place.

The appointment of an individual to act as a "protocol enforcer" is necessary to ensure the grounding of the learning into day-to-day operations. This individual assists the team in its daily functioning by monitoring individuals' activities as well as team meetings and project reviews. He gives feedback to the individuals and to the team and may directly participate in operational activities by providing examples of best practices. The "protocol enforcer" organizes short wrap-up seminars to review the implementation of the protocols learnt in each module and their impact on team's performance.

2.3.1 Context's Assessment

The stake for context's assessment is validating the feasibility of the implementation of the M3C. The context's assessment brings awareness to the individuals of the need for and rationale for change. The goals of the CA module and the methods to carry it out are presented in Table 2.3.

Participants are asked to fill in the questionnaire a few days before the first seminar and then regularly at the beginning and at the end of each phase. This recording of perceived status permits the monitoring of the evolution of the perception of people and therefore the climate in the team.

The context's assessment module should be implemented one to three weeks before the beginning of the community norming module to maximize impact on the learning process. During the interviews, the facilitator makes the stake, the aim, and the goals of the change process explicit and creates a triggering event useful to start the unfreeze phase of the change process (Lewin, 1950, 1951). The interviews acquaint individuals with the facilitator and help in creating interpersonal relationships that will facilitate the learning processes in the coming seminars. By observing the reaction of the individuals, the facilitator can identify who will be resistant to change and who might act as a resource. The facilitator can send individuals customized messages to be digested in the waiting period. Such messages are aimed to create awareness of specific behaviors that could help or jeopardize the change process in the community.

Goals
- To understand the perception that the driver has of the context - To check his willingness to support the complete implementation of the model - To define a list of items to be measured and monitor the context and assess progress
- To map the perception that individuals, directly or indirectly, affected by the change initiative have of the context
- To verify, cross-check, and validate the consistency of the perceptions of the individuals involved in the change process - To make the driver aware of the differences between her perception and the perceptions of her colleagues - To confirm driver's willingness to implement the model
- To tailor-make the implementation of the model - To agree on the conflict cleansing process (if needed)
Methods
- To carry out semi-directive interviews by using the FiP (Fixed Points) and SAS (Slots of Active Support) protocols (see part three) - To co-design with the driver the follow-up questionnaire
- To carry out non-directive interviews with all the individuals involved in the change process (or at least with a representative sample of them if they are too many) - To administer the follow-up questionnaire
- To consolidate and to make anonymous the result of the interviews and of the follow up questionnaire - To report it to the driver and to discuss her reaction
- To co-design with the driver the training program (activities, pedagogical approach, schedule, participants, etc...) - To set feasible objectives for the learning process to be announced in the community norming module

Table 2.3—Implementing the context's assessment module

It may happen that during the interviews, individuals make reference to hidden or explicit issues that they face in dealing with other members of the team. The facilitator has to determine if the issue is a real problem concerning an interpersonal conflict, or if the individual is just experiencing a difficulty in dealing with the others (Watzlawick, Weakland, and Fisch, 1974).

When an individual is experiencing difficulty, he autonomously uses his energies to look for solutions that involve the others. The individual may be open to alternative solutions. In this case, the role of the facilitator is to help the individual become conscious of the present behavior(s) and/or perceptions that resulted in the difficulty. The individual may experience an insight to identify new solutions to the issue before it drifts into a more dangerous conflict.

If the individual discovers corrective actions through insight, the probability of successful implementation increases (Lax and Sebenius, 1986, de Saint Paul, and Tenenbaum, 2002), and the probability of escalating interpersonal conflicts is reduced.

When an individual is confronted with a problem (e.g., a conflict involving the inappropriate behaviors and/or attitudes of others), he cannot autonomously provide acceptable solutions. The individual's resistance to changing only his behaviors/attitude increases. The individual focuses his energies on protecting himself from others' attacks or attacking others.

When the facilitator detects a problem, he should recommend the driver implement the conflict cleansing process. In the conflict cleansing process, the facilitator uses the Emotional Leveling protocol (see part three) to help the individuals involved in the conflict reduce or eliminate interpersonal or inter-organizational blocking issues and concentrate their energies on the change process.

Conflict cleansing may appear to the driver as a long, difficult, dangerous, oversized, useless, and out-of-scope process. However, it is one of the key factors for success. The successful implementation of any strategy and the effectiveness of any structure rely on the commitment, emotional as well as intellectual, of the individuals affected by the change initiative. The organization's performance is determined by its members' capability of understanding, creating, and supporting interdependence among them. Emotional conflicts distort reasoning, objectivity, and logic, and if ignored or placated, become even more distorted, pervasive, and resistant. Interpersonal relationships constitute powerful meta-communication channels, and meta-communication affects any type of relationships (Watzalwick, Beavin, and Jackson, 1967; Watzlawick, Weakland, and Fisch, 1974). Any blocking factor in individuals' meta-communication may jeopardize the implementation of any strategy or the functioning of a structure and put the stake in danger.

The emotional leveling protocol facilitates communication between individuals or teams involved in a conflict because it clarifies the emotions driving the inappropriate behaviors. Emotional leveling can be used either by an external mediator or by the very same people concerned by the conflict (Rosenberg, 1999).

2.3.2 Community Norming

The stake for community norming is setting the team responsible for implementing the M3C. The CN module focuses on creating the cohesiveness of membership to enhance the team's stability during the change process.

Table 2.4 presents the goals of the community norming module and the methods to carry it out.

All the activities performed during the community norming module are based on intensive individual and/or collective work, strong participation, and high participant involvement. For the learning process to be successful and effective, the facilitator must ensure psychological and emotional safety to reduce stress and learning anxiety (Schein, 2002).

Experiences related in the case studies (part four) shows that the outcomes observed after the implementation of the community norming module are:

- Improved interpersonal relationships between the participants;
- Better cohesiveness of the team and increased cooperation amongst individuals and departments;
- Stronger commitment to the change process and support to the driver in the implementation of the M3C;
- Increase in teamwork productivity.

	Methods
- To set-up ground rules to facilitate cooperative teamwork.	- To make individuals playing the "prisoner dilemma" show the usefulness of practicing the cooperation emergence protocol (Axelrod, 1984);
- To put together divergent interests to protect the identified common stake.	- To make individuals participating in games and role plays learn and practice the creative negotiation grid and the power patterns matrix. - To explain the feedback matrix and to make individuals practice it immediately.
- To establish a consensual decision-making process to avoid implementation failures. - To make individual and collective commitments explicit to ensure long-term support to action plans.	- To make individuals playing games and simulations to practice various decision making processes and test their appropriateness to the different contexts (see part three) - To make individuals accustomed to use the option-based decision-making and strong consensus protocol as a general manner of behaving in implementing the action plans.

Table 2.4—Implementing the community norming module

2.3.3 Contextual Drivership

The stake for contextual drivership is learning and co-opting the drivership tools used to initiate and regulate actions. The CD module is to ensure individuals' powerfulness to enhance team's performance. Table 2.5 presents the goals of the contextual drivership module and the methods to carry it out.

Experiences related in the case studies (part four) show that after the implementation of the contextual drivership module individuals know how to:

- Structure the relationships among the elements of the system by making explicit goals, roles, responsibilities and means (Van Krogh and Grant);

- Identify their own and colleagues' level of reliability;
- Agree on the level of autonomy of the contractual relationship in order to adapt reciprocal behaviors;
- Support and foster own and colleagues' commitment and competencies;
- Follow up colleagues' actions and support them when necessary.

Goals	Methods
- To fine-tune system's functioning, interfaces' quality, and activities' consistency to organize the complexity of interdependent actions.	- To make the client cascading the stake into sound and consistent individual/collective objectives by making the organizational alignment map explicit; - To make each individual filling in his driver's added value matrix to learn how to set the right mix between actions planning and tasks allocation; - To make each individual checking the consistency of his own driver's added value matrix with those of the other colleagues involved in the interdependent actions;
- To support commitment and to develop competencies by making relationship rules explicit. - To facilitate appropriate feedback at any level to adapt team's actions by correcting deviations and errors.	- To use PBL (Problem-Based Learning) to make individuals learning the protocols of contextual drivership; - To make the team practicing according to the following steps: 1. Individuals apply the protocols in role plays, simulations and forum-theatre; 2. The facilitator helps each individual to apply the protocols to himself; 3. The individuals meet in pairs and use the protocols to analyze their operational relationships, with the assistance of the facilitator when needed; 4. The team explicitly commits to adopt these protocols and define an action plan to reshape the functioning of the interfaces.

Table 2.5—Implementing the contextual drivership module

2.3.4 Cooperative Innovation

The stake for the cooperative innovation module is producing meaningful solutions. The activities of the CI module stimulate creativity to enhance team's competitiveness.

Table 2.6 presents the goals of the cooperative innovation module and the methods to carry it out.

Goals	Methods
- To create and to support the reflex of positive contribution by focusing on solutions instead of problems - To concentrate resources to tackle issues from different points of view in a cooperative way - To produce efficient, effective, simple, feasible solutions to complex issues - To encourage forward thinking, knowledge sharing, and best practices seeding - To implement innovative work team technologies and processes	- De Bono's parallel thinking (1985) - Groups for exchanging and progressing - Benchmarking and best practices

Table 2.6—Implementing the cooperative innovation module

PART THREE: PROTOCOLS TO IMPLEMENT THE MODEL OF COOPERATIVE CONTEXTUAL CHANGE

3.1 CONTEXT'S ASSESSMENT

The purpose of the context's assessment module is to provide the driver and his colleagues with a set of processes, tools, and techniques to bring awareness of the need for change. Such awareness is a pre-requisite for the applicability of the model.

The context's assessment module must be designed to clearly identify the stake of the change process, to make sure that such stake is understood and accepted at any level in the organization. This allows then to align organizational goals at any level, ensuring the consistency of people's efforts during the change process.

The change process triggers fears, frictions, and blockages, both at the interpersonal and inter-organizational level. The context's assessment module must then allow for specific actions to reduce such resistance.

It is then necessary for the driver to identify the appropriate tools and criteria to measure the effectiveness of the change process, both at the emotional and operational level, in order to monitor the implementation and take the required corrective actions.

3.1.1 Emotional Leveling Communication Protocol

Table 3.1.1 describes the emotional leveling communication protocol, which is an adaptation of the non-violent communication protocol presented by Rosenberg (1999). It can be used to mediate between two counterparts (individuals or groups; in the latter case a representative will speak on behalf of the group).

Phase	Rationale	Protocol's steps and actions carried out by the individual
1: context analysis and problem formulation	The aim is to help people in identifying what concerns the content level (actual issues on which people disagree) and what concerns the meta-communicational level (relationship conflict because of emotional issues) (Watzlawick, Helmick, Beavin, Jackson, 1967).	*Step 1 - Observing my reality* I factually list the events that I observed without any reference to my impressions or opinions
	Separating facts from feelings helps to separate people from problems, in order to be "*hard on problems and soft on people*" (Fisher, Ury and Patton, 1981, 1991)	*Step 2 - Recognizing my feelings* I express what I feel because of these facts by making reference to my own emotions
2: relationship setting and solutions seeking	The aim is to help people in agreeing on the meta-communicational level (relationship features), in order to concentrate their energies on solving then the disagreement at the content level (Watzlawick, Helmick, Beavin, Jackson, 1967).	*Step 3 - Acting first to improve the relationship* I make explicit which specific actions I commit to take in order to improve my relationship to the other person
	Committing to be the first in acting to improve the relationship and being explicit on what the other person is asked to do helps in identifying what actions could contribute to a mutually beneficial conflict settlement.	*Step 4 - Telling my needs and asking for reciprocity* I make explicit what I need the other person do to better bear such feelings and emotions and go on in the relationship with the other person

Table 3.1.1—Emotional leveling communication protocol

From an operational point of view, the proper implementation of the emotional leveling communication protocol requires that:

- The facilitator explains the protocol to the two parties and makes sure that they both agree on going on with the process. If one of them does not agree, the facilitator may decide to let the individual who agrees to carry his own part of the protocol while asking the individual who disagrees to just listen to his counterpart. Tossing a coin is the recommended way to decide who will speak first.

- The two individuals (let's call them A and B) are given no more than seven minutes to prepare what they will say to the counterpart.

- Individual A goes through the process by clearly separating each step in the process: he has no longer than seven minutes to read what he wrote in step two; when individual A speaks, individual B takes notes on what individual A is saying, keeps silent, and does not interrupt (but to ask individual A to slow down or repeat a sentence).

- Once individual A has finished speaking, the facilitator asks for a short silence break. If necessary, and only in case of deep interpersonal issues, the facilitator may ask individual B to read what he wrote while person A was speaking, starting each sentence with "what I heard you say was ..." Individual B quotes as exactly as possible (the facilitator helps clarify as appropriate, as explained later) what the other person said.

- The same process (steps two to four) is repeated but with inversed roles: individual B speaks while individual A takes notes. B has to read what he wrote in step two without adding anything related to what was said by A in step three.

- The facilitator asks the two parties if they want go on with another round of the process. In this case, the facilitator will start again from step one.

- If the two parties agree on closing the process, the facilitator reformulates the commitments and wishes of both parties before letting them go.

The facilitator must ensure that the individual uses the first person when speaking, starting his sentences with I. If the speaker uses the second person (you), the facilitator asks him to rephrase the sentence (or he will rephrase if needed).

People, once trained in the use of the protocol, are encouraged to use it even without the help of the facilitator.

The emotional leveling communication protocol can be as well used to help two groups to communicate when facing a conflicting issue. In this case, each group appoints a speaker, who is the only member of the team allowed to speak during the protocol, who reports what has been decided collectively by the team.

The effectiveness of the emotional leveling communication protocol is due to the fact that the process is:

- Short: both individuals have only seven minutes to prepare what they want to say and seven minutes to effectively tell it. Time restriction pushes them to focus on the first things that come to their mind, which tend to be the most important.

- Asynchronous: when an individual speaks, the other keeps silent. Since the individuals prepare before speaking (and listen to the other's arguments), it is not possible for the two parties to enter endless polemics or sterile discussions. Both individuals must concentrate their energies on listening to the other's point of view instead of using them to find counterarguments while the other is speaking (Watzlawick, Beavin, and Jackson, 1967);

- Synaesthetic: Participants are obliged to use their three sensory channels: visual (when reading their notes or quoting the other person's issues), auditory (when listening to the other's arguments or the other person's quoting), and kinesthetic (when writing their own arguments or writing the other person's arguments). As suggested by the neuro-linguistic programming model (de Saint Paul and Tenenbaum, 2002), this process optimizes the chances the other's message will be received, understood, and acknowledged at the highest degree and with minimal distortion.

These three features are sometimes seen as obstacles to the "free flow" of communication and settlement seeking. Paradoxically, they allow the individual to better understand the other's premise and perception of the issue. Mutual understanding encourages resolving the relationship part of the conflict before concentrating on the content part of the conflict (Watzlawick, Beavin, and Jackson, 1967).

The emotional leveling communication protocol enhances the exchange between the two parties because it grants each individual the opportunity of telling freely what is important for him and the assurance of having been listened by the counterpart. In fact:

- During the preparation phase, each party does not know the other party's argument; this prevents them from responding immediately and eliciting an emotional defense;

- The one speaking must always use first person, cannot directly blame the other person (which invariably arouses emotional reaction), and may only reference observed facts;

- No agreement is required on a joint action plan; instead each individual commits to unilateral actions.

3.1.2 Strategies for Reducing Actors' Resistance to Change

Resistance to change can occur at three levels: individual, group, and organizational. The change process requires the alteration of traditional norms of functioning and the modification of associated behaviors. At the same time, the new norms may call for the development of new values, which may diverge from the existing ones (Soparnot, 2003).

To succeed in the implementation of a change process, the group must explicitly have a common language and share basic assumptions (Schein, 1985). The group must be able to translate this language and assumptions into norms, values, and ground rules in order to establish new, formal, and/or informal contractual relationships, which bind the individuals together (Trice and Beyer, 1993).

The first two steps of the model of cooperative contextual change require the driver to:

- Present what appears to be the explicit stake, elicit the stake perceived by the actors involved in the change process, and resolve any differences;
- Allow people to co-build and co-opt new ground rules to appropriately handle the interfaces necessary to protect the stake.

Discussion of the stake is necessary because it impacts the values of the individuals, groups, and organizations. The stake references the ideals people share and serves as a moral benchmark against which aspirations and behaviors are judged. Values define what is "good" and what is "bad" and provide a rationale for participation in collective action (Durkheim, 1893, 1984; Morrison, 1995).

Values generate the norms and rules, which guide the functioning of the group and the relationships among individuals. Norms and rules impact the pragmatic way of life by shaping what individuals see as being the "right" and "wrong" way to implement change. Values, norms, and rules provide people with the feeling of security (Durkheim, 1893, 1984; Morrison, 1995).

At the cognitive level, ground rules shape solution-finding in regard to relating with each other and dealing with the external environment. At the emotional level, ground rules clarify member integration and the issues related to authority and influence.

Ground rules facilitate the creation of solidarity and the dealing with conflicts (Schein, 1985).

Discussion of the stake may influence individuals to accept or resist the change process (Argyris, 1995; Soparnot, 2003). Crossing the two dimensions "opinion on the stake of the change" and "opinion on the change process implementation" produces the actors' behaviors matrix, presented in Table 3.1.2. This matrix provides a behavioral taxonomy that the actors may exhibit with regard to the change process. It facilitates the designing of appropriate relationship strategies with each actor. The driver may customize the relationship for a specific type of actor so as to create conditions, which facilitate the actor's willingness to become involved in change (Bovey and Hede, 2001a; Maurer, 1996c; Waddel, 1998; Erickson, 1967, 1980, 1992; Haley, 1973, 1976; Watzlawick, Weakland, and Fisch, 1974; Watzlawick, 1983).

		Opinion on the change process implementation (pragmatic appraisal, with reference to norms and rules)	
		Wrong	Right
Opinion on the stake of the change (moral judgment, with reference to values)	Good	*Critics* While adhering to the stake of the change, they think that the change should be implemented in a different way. They may or may not actively provide suggestions to modify the change process	*Change Agents* They adhere to the stake of the change and agree on the way the change is implemented. They can either behave proactively to promote it, or quietly support in the background
	Bad	*Opponents* Do not adhere to the stake of the change and do not agree with the way change is implemented. They may or may not make explicit their opposition.	*Followers* They do not adhere to the stake of the change (either because they do not agree with it, or because they did not understand it). However, they implement change, either passively or reacting when required

Table 3.1.2—Actor's behaviors matrix

With regard to change agents, the driver must establish conditions which support their commitment, protect them when needed, and provide resources to carry out the actions necessary to implement change. The driver must respect the change agents' action preference, to act either publicly or provide hidden support. When the change agent produces positive outcomes, the driver must ensure that the other actors understand how these results contribute to the welfare of the community. People who understand the benefits of involvement are more likely to join the change agent's team.

However, the action of the change agents can sometimes be disruptive, especially when pushed by the passion. If they fail to perceive and manage resistance or if they do not acknowledge the good reasons that others may have to see things differently, inadvertently they create the conditions for a coercive implementation of change (Maurer, 1996c).

With regard to the critics, the driver and the change agents must actively solicit their suggestions, comments, and ideas. If their contributions are appropriate, they should be acknowledged and incorporated. Critics who are asked to be involved in the change process are more likely to consider it as the "right" process because of their contribution. Forcing critics to comply may increase resistance and transform them into opponents (Maurer, 1996a).

Followers' behaviors may range from malicious compliance to easy agreement. In the former case they say they agree with the change but covertly undermine it. In the latter case, individuals may agree with the way the change is implemented but are unwilling to put time and efforts to cope with the effects of change (Maurer, 1996a). The driver and the change agents must provide followers with examples of successes (quick wins), which can then fuel their involvement and increase their commitment. At the same time, drivers and change agents must be willing to continuously stimulate followers to make them intimately accepting the change.

Understanding and working with opponents is one of the most sensitive issues. Opponents' resistance may range from explicit opposition to denial that the problem exists (Maurer, 1996a). The driver and change agents may ignore their vocalizations and divisive actions, concerned that control will be lost. However, if opponents are not allowed to express their opinions, act publicly, or their ideas are rejected without proper consideration, they could cast themselves as martyrs. In this case, followers or critics could feel sympathetic and join them. The driver and change agents must act to avoid this consequence by recognizing opponents useful role and co-opting the negativity they can have on the others (Watzalawick, Beavin, and Jackson, 1967; Watzlawick, Weakland, and Fisch, 1974; Haley, 1973, 1976). This can be accomplished by establishing clear rules of functioning, which permit the expression of their ideas while controlling rumor, innuendo, and "false facts." Providing factual information prevents opponents from leveraging the lack of transparency. The cooperation emergence protocol (Axelrod, 1984), presented in chapter 3.2, is recommended as a way to establish such rules. In general, shifting the discussion to the level of belief and identity rather than at the level of norms or values can prove to be successful in overcoming differences in dealing with the opponents (Bovey and Hede, 2001b). Thus the discussion centers on the stake and its impact on the life of all the community.

3.2 COMMUNITY NORMING

The purpose of the community norming module is to provide the driver and his colleagues with a set of processes, tools, and techniques to create, protect, and develop a team's cohesiveness and effectiveness. Understanding the importance of adopting a cooperative approach and using sound decision-making protocols helps the group successfully implement any change process.

The community norming module must be designed so that the group learns how to co-build and co-opt the ground rules and the operating modes, which will allow them to find the most appropriate ways of functioning to protect the stake.

At the beginning of the first seminar, it is helpful to introduce the concept of ground rules and operating modes and show how to build and to enforce them in a cooperative way by having group's members:

- Play the prisoner's dilemma to explain and co-opt the cooperation emergence protocol;
- Self-define the basic ground rules, which will regulate the communication and participation issues during the following seminars.

These exercises provide an opportunity to explain and co-opt the decision-making and strong consensus protocol, and to have the group using it to co-opt the ground rules.

Such co-opted protocols and ground rules are enforced during the various seminars to provide people with a living example of how using a simple principle generates a long-lasting quality context. In the periods between the seminars, the "protocol enforcer" must encourage people to use these operating rules and protocols in their day-to-day professional relationships. Practicing "new ways of relating with others" is likely to provide the "quick wins," which reinforce people's keenness to implement and enforce the ground rules.

3.2.1 Cooperation Emergence Protocol

The cooperation emergence protocol (Diagram 3.2.1), inspired by Axelrod (1984) grounding research on people's behaviors in playing the Prisoner's Dilemma[1], helps to understand why a strategy of relationship based on reciprocity can facilitate the emergence of cooperation. To successfully promote cooperation with other actors, a relationship strategy must be:

- Respectful (it never defects first, which prevents from getting into unnecessary trouble);
- Reactive (it replies immediately to other's defection, but only at a nine-tenths level, which discourages the other side from persisting whenever defection is tried);
- Forgiving (it goes back to cooperation as soon as possible, which helps restore mutual cooperation and avoid escalating the conflict);
- Clear (it is made explicit and explained, which makes behavioral pattern easy to recognize.

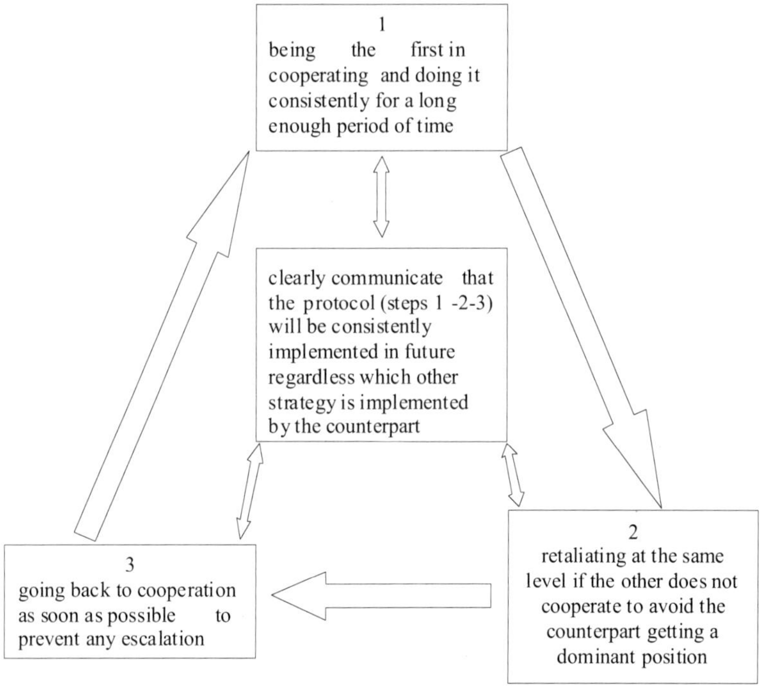

Diagram 3.2.1—The cooperation emergence protocol

The focus of optimal forgiveness depends upon the context. If the probability of unending mutual recriminations is high, then a generous level of forgiveness is appropriate to break the cycle. When facing strategies that are good at exploiting easygoing rules, an excess of forgiveness is costly. A one-for-one response (reciprocate denial of cooperation as well as cooperation) is likely to be quite effective in a wide range of settings.

Axelrod's research (1984) shows that such a strategy is

- Robust; that is it can thrive in a variegated environment composed of others using a wide variety of more or less sophisticated strategies;

- Stable; that is it can resist invasion by mutant strategies;
- Initially viable; that is it can get a foothold in a context that is predominantly non-cooperative.

3.2.2 Creative Negotiation Grid

In the process of defining common ground rules and operating modes to protect a common stake, team members must continuously negotiate trade-offs regarding divergent interests internally and externally.

The creative negotiation grid (Table 3.2.1) integrates the principled negotiation method developed at the Harvard School of Law Program on Negotiation (Fisher and Ury, 1981, 1991; Lax and Sebenius, 1986, Ury, 1991; Shell, 1999) with systemic communication techniques like neuro-linguistic programming and transactional analysis (Cialdini, 2001; de Saint Paul and Tenenbaum, 2002; Tenenbaum, 2002). Such an approach can provide a roadmap, which helps people align efforts to protect their common stake while satisfying, as much as possible, their individual interests.

Five elements	Four phases	Three key moments	Two factors
the stake the prioritized interests the BATNA and the other alternatives the legitimate criteria the creative options	Analyzing the context and preparing the negotiation Sharing information Trading off interests and bargaining Implementing the agreement	Anchoring Suggesting trade-offs Proposing the deal	the context the negotiators

Table 3.2.1—The creative negotiation grid

An elaboration of the creative negotiation grid, which puts into practice the principles of the creative negotiation grid, follows:

The Five Elements of a Creative Negotiation

The five elements of a creative negotiation consist of: (1) the stake and (2) the prioritized interests, (3) the BATNA and other alternatives, (4) the legitimate criteria, and (5) the creative options.

The Stake and the Prioritized Interests

The stake is the most valuable assets, tangible or intangible, that the driver (or the community) wants to protect at any cost. At any decision level, the driver or the community seeks to ensure that the result of a decision set and associated actions protects or enhances the stake. The stake is the persistent reference that defines the direction and orientation of any action, and if the contribution is useful in protecting the stake. The stake is not negotiable and cannot be traded off. However, if the context changes, it may be necessary to modify or completely change the stake.

Once the stake has been identified, the driver must prioritize the various interests that he and his colleagues want to satisfy within the change process. Prioritizing interests establishes the conditions of a cooperative approach to value creation (Lax and Sebenius, 1986; Ury, 1991) instead of focusing on individualized positions. Making clear the most important, valued interests to be satisfied helps:

- Identify the stake to be protected and defines the goals to be met;
- Choose which interests that can be traded off, the least important, when it is necessary to make a concession. Clearly identifying the stake helps the driver build and communicate to his colleagues the vision to be realized and establish a stable direction during the change process.

The implementation of the contextual drivership method is based on the identification of the driver's stake (and prioritized interests) to ensure the organizational alignment chain and to properly use the autonomy curve.

The BATNA and Other Alternatives

A driver may be in a position of having an action or set of actions that can be carried out alone or with other parties external to the negotiation, which protect his stake better than negotiating with his colleague. Having an alternative action to negotiation frees the driver from having any link with the incumbent colleague.

Being able to build alternatives to negotiating is one of the most important skills of a driver, because it allows the designing of scenarios activities (see the driver's Added Value matrix).

The concept, Best Alternative to Negotiated Agreement, (BATNA) can be a key factor of stability in a change process (Fisher, Ury, and Patton, 1981, 1991; Lax and Sebenius, 1986; Ury, 1991). The BATNA allows the driver to enter the negotiation process with a point of reference, to which the value of the agreement can be compared, allowing him to be more realistic about his own case. If the driver has a weak BATNA, he may be more willing to accept tough conditions imposed by his colleague(s).

A strong BATNA increases the driver's negotiating power, because it may be used as a backup if the negotiation process turns difficult. In general, if the driver has several alternatives, he will be able to better deal with objections from his colleague without taking a defensive position. This will increase his stability, his credibility, and his negotiation power (Rogers, 2003).

Since change is not a deterministic process, the individuals driving the initiative or response must be flexible and ready to adapt to any alterations of the environment and context.

The Legitimate Criteria

The driver, during the change process, provides True, Relevant, Understandable, and Evoking Interest (TRUE) facts to support his statements. When facing objections from colleagues, the driver must use legitimate criteria, not take a defensive position, and focus on helping colleagues become aware of the need for change (Rogers, 2003).

The Creative Options

An objective for the driver is to stimulate colleagues to seek creative options, when facing an issue, and develop a solution-oriented mindset. Conjoint development of alternatives reduces the risk of unnecessary conflicts and fosters team stability in dealing with difficulties and problems. Identifying several options helps the driver suggest trade-offs and enhances agreement. The deeper the preparation, the wider the exchange of information, the clearer the stake/interests of each party, and the easier the development of options. Options are a concrete illustration of the variety of outcomes that satisfy joint interests (Ury, 1991). A skilled driver encourages and facilitates colleagues taking responsibility for identifying the need for change, the existing obstacles, and the applicable solutions, which foster the change process. In addition, the driver expects colleagues to protect individual and common stakes while satisfying, as much as possible, the interests of each party (Rogers, 2003).

The Four Phases of Creative Negotiation

Analyzing the Context and Preparing the Negotiation

To achieve an effective and successful negotiation, the driver must have sufficient breadth and depth of knowledge regarding the features of the specific context in which the change process takes place to articulate the main benefits that the change process will bring to the community (Rogers, 2003).

Any negotiation process starts with a comprehensive and extensive analysis of the five elements of the creative negotiation grid. The driver has to establish a condition of analysis sufficient to determine the five elements. The preparation sheet (Table 3.2.2) articulates the five elements of the negotiation relevant to the driver, to colleague(s), and any other party directly or indirectly involved in the change process. Completing the grid creates the condition for reflective thinking, which helps the driver identify the most suitable strategy and tactics to be used in the next steps. Failing to prepare is preparing to fail.

Elements	Party One (myself)	Party Two	......	Party n
The stake				
The prioritized interests				
BATNA and other alternatives				
The legitimate criteria				
The creative options				

Table 3.2.2—The creative negotiation preparation sheet

Sharing Information

Sharing information is particularly important in any negotiation process. The sharing process allows the parties to create, develop, or strengthen their interpersonal relationship and understand their respective interests and needs. Sharing tests colleagues' commitment and willingness to agree on specific criteria to be used to making a decision. Options that can create value for all the parties involved in the process and possible obstacles to the agreement can be identified and discussed (Shell, 1999).

The driver asks questions designed to reveal colleagues' position on the issue(s), carefully analyses the position statements, and leverages these statements as barriers to prevent colleagues from retreating from an accepted or agreed upon point (Rogers, 2003).

The driver strives to link the benefits of the change process to the needs and beliefs of the colleague, which creates the conditions for a successful agreement. The driver may be able to use the openness of the sharing information phase to leverage the colleague's needs to reduce dissonance, which can facilitate the coming decision-making process (Shell, 1999, Rogers, 2003).

Trading off Interests and Bargaining

As information is shared and positions become known, the need to trade off interests and bargain to retain as much of one's position as possible begins. This phase is usually the most difficult for the driver, because colleagues are more likely to raise objections, perceived, real, or pretend. Initially, the driver treats every objection as valid and with merit. The driver listens carefully to the colleague and seeks some point, which both parties can agree upon as a basis to overcome the objection. In any case, he never argues in handling objections. Rather, suggestions are made to help the colleague discover a satisfactory answer to her/his objection (Rogers, 2003).

If a pretended objection is presented, handling it as true calls the colleague's bluff, because resolving a pretend objection commits the colleague. If the objection is deemed irrelevant, the driver should ignore it. When an objection is true and relevant, it is important that the driver does not react immediately to it: he should think about and then respond to it in a reasonable time. If a valid objection is unanswered, it becomes an obstacle to the agreement (Rogers, 2003).

Another important element of this phase is the capability of the driver to deny without denying directly (Ury, 1991). Indirect denial offers acknowledgment of the other's position, while presenting the position of the driver without rancor. Denying using a question is as well very effective, because it forces the colleague to state by himself the denial (Rogers, 2003).

Implementing the Agreement

The negotiation process is finished only when all the elements of the agreement are identified and definite actions formalized by the driver and colleagues. The more complex the issue being negotiated, the more critical is agreeing on the implementation process and activities. At this juncture, identifying and committing to the activities, responsibilities, and milestones within a timeframe is essential.

The Three Key Elements of Creative Negotiation

Anchoring

The driver must create conditions favorable to negotiating from his position. Anchoring involves establishing the pitch of the process (conflict or cooperation, coercion or participation) as well as the principles, the ground rules, and the criteria that will regulate the process. Being the first to anchor does not mean speaking or opening first. Anchoring attempts to create, through ethical means, the framework of reference, which guides the negotiating process. Mastering the opening techniques and knowing perfectly the context of the negotiation are necessary conditions for a successful first move (Shell, 1999; Rogers, 2003).

The stronger the anchoring, the sounder the negotiating position. Just as in chess, the driver should master several openings, which permit adapting to various contexts and increase the probability of being able to control the board—anchor first. When the negotiation involves a complex issue, the process may require successive mini-negotiations, each with its own opening. Prescribing the symptom in the anchoring phase can be very effective to reduce some of the resistances to change (Erickson, 1967, 1980, 1992; Haley, 1973, 1976; Watzlawick, Weakland, and Fisch, 1974).

Suggesting Trade-offs

If the driver is the first in identifying possible reciprocal concessions, he can present less important interests and conduct the trading off process in the most favorable position. Initiating trade-offs may provoke the "reciprocity reflex" in the colleague (Cialdini, 2001), which may reduce colleagues' reluctance and help reduce the need to raise objections. Secondarily, driver initiated trade-offs may allow the colleague to "save face" and encourage him to accept the conditions that are important for the driver.

The driver may be able to sustain his controlling position by eliciting the contrast principle and the norm of reciprocity (Cialdini, 2001; Shell, 1999) through:

- Anchoring high;
- Proposing a set of elements that most fit with the legitimate criteria used by the colleague in assessing the offer;
- Making a few minor and slow concessions.

Proposing the Deal

Just as with trade-offs, being the first to offer a deal setting may reduce colleagues' anxiety, since a solution to resolve the issue is offered (Lax and Sebenius, 1986). Solution initiating may facilitate shaping the agreement implementation plan around the points important to the driver. Initiating a solution signals that the trade-off and bargaining phase is over, the time remaining is short, and implementation needs to begin. This promotes the "scarcity effect" (Shell, 1999; Cialdini, 2001). The proffered solution should include elements that will help the colleague reduce future dissonance (Shell, 1999, Rogers, 2003). In proposing a closing agreement, the driver should establish conditions that encourage the colleague not to try leveraging the proposal to open a new bargaining phase (Rogers, 2003).

The Two Factors of a Creative Negotiation

The Context

Every negotiation is unique because its context is unique. Each context is unique because of the stake, relevant interests of each party, and the decision-making process used by the parties. The driver uses the preparation and the information-sharing phases to properly analyze the context, shape his own understanding of it, and adapt his behavior accordingly. The uniqueness of context prevents a universal recipe for a successful negotiation. But, as indicated above, guidelines for increasing the probability of a successful negotiating are available.

The Negotiators

The negotiators, their characters, their negotiating styles, their background, the type of relationship they had in the past, or are likely to have in the future, are unquantifiable factors that influence a negotiating event. Because negotiating involves working with individuals to build and support a stable relationship and extensive training, focusing on understanding that negotiating means playing a piece of theatre is necessary. Negotiating sometimes may oblige participants to play roles that they did not choose or like, roles that require acting differently from what they would do with forethought or spontaneously.

The knowledge the driver has of himself, his capability of having a distant view of the negotiation context, and his mastery of self are critical elements in negotiating (Ury, 1991). Rogers (2003) suggests that negotiating with difficult people—conceited, cynical, antagonistic, self-important, argumentative, irritable, or close-minded—requires the following:

- Be patient, calm, cool, firm, and treat the colleague pectfully and courteously at all times. Use understatement wherever possible;

- Proceed slowly and with care. Listen attentively. Use plenty of questions. Where possible, repeat statements made by the colleague before dealing with them. Give proof for every point made;

- Note carefully the objections being made and determine whether these are "real" or just "good" reasons. Deal with the real objections in detail but try to let the colleague talk;

- Don't become involved in argument but suggest possibilities.

In essence, a good change agent reveals wisdom in using his knowledge of the context of the negotiation and an enthusiasm for the content of it.

3.2.3 Feedback Matrix

A system is composed of interdependent parts. Any change concerning one part affects (directly or indirectly, at a lower or higher level) all the other parts. The stability of a system, as a whole and each of its parts depends on the performance of the processes, which regulate the interactions and interfaces between and among the parts (Grant, 2001).

The emotional leveling communication protocol, presented in part three, chapter one, is designed to manage conflicts. The feedback matrix (Table 3.2.3) is used to avoid the drifting of the relationship towards a conflict. The feedback matrix provides positive reinforcement for the desired, virtuous behaviors, and corrective hints when mistakes occur. Its frequent use facilitates the successful implementation of the model of cooperative contextual change, because it contributes to reduce resistance to change (Anderson and Ackerman Anderson, 2001).

Appropriate feedback has the following features (Blake and Mouton, 1985; Watzlawick, Weakland and Fisch, 1974):

- It constitutes a non-assessment and non-judgmental description of facts;
- It is specific and focuses on major issues and not on minor details;
- It addresses observable behaviors that the person can change instead of looking at inner attitudes that might be out of his direct control;
- It is given with the purpose of helping the recipient to improve and not to head a fault or to make the person feeling guilty.

The feedback matrix is the combination of two dimensions (the focus and the purpose of the feedback) and helps to visualize four possible types of feedback shown in Table 3.2.3.

The focus of the feedback can be on:

- State of acting, the observed actions, behaviors, results, and comments that the recipient has done or achieved;
- State of being, the personal, physical, emotional, psychological, and intellectual qualities, characteristics, and attitudes of the recipient;

The purpose of the feedback can be:

- Reinforcing makes explicit that the provider recognizes and approves recipient's actions, behaviors, qualities, and attitudes;

- Reorienting makes explicit what should be changed in the recipient's actions or behaviors to fulfill provider requirements. To provoke effective change and prevent uncontrolled drift in relationships, it is important to focus the feedback on what can be done to improve the situation, instead of focusing on what is wrong and why it happened (Watzlawick, Weakland, and Fisch, 1974). Reorienting feedback on the state of being should be avoided.

A driver can leverage three types of feedback to maintain or improve a relationship, which creates, supports, and develops people's commitment. (See commitment enhancement matrix). Public reinforcing feedback on the state of acting is a healthy practice. Reinforcing feedback on the state of being and reorienting feedback on acting should be done in private.

The feedback protocol must be used extensively to have its greatest effect. Therefore, a majority of influential individuals, within the organization, must agree on the following rights and duties (Tenenbaum, 2000).

- To give feedback—Every individual is responsible for giving feedback to the others (regardless of the hierarchical status), when he thinks it is necessary to consolidate the actual relationship or to avoid its drifting toward conflict. The provider must give feedback in the appropriate way (as shown in the feedback matrix) for the feedback to be acceptable by the recipient;

- To ask feedback—Every individual can ask for feedback to the others, when he thinks that the others can give him hints on how he should reconsider his premises and modify his relationship with the colleagues. The driver should be the first to ask for feedback, to give the example,

		Focus of the feedback	
		Acting	Being
Purpose of the feed-back	Reinforcing	*Yes* it is useful to support recipient's virtuous behaviors it strengthens recipient's self-confidence in his resources and competencies it reduces recipient's stress and fuels his commitment	*Yes* (if accepted by the recipient) it is useful to strengthen recipient's self-confidence in his personal qualities it may impacts inner motivational factors it increases recipient's well-being and self-confidence
	Reorienting	*Yes* it must be factual, specific, focused on concrete corrective actions it is useful to stop undesired behaviors and to provide new patterns of action it helps the recipient to correct occurred mistakes and to avoid repeating them in the future	*To Be Avoided* it concerns what is related to the inner world of the recipient (attitudes, feelings, emotions) it belongs to the realm of therapy demanded by the recipient

Table 3.2.3—The feedback matrix

to promote the implementation of such practice, and to help his colleagues to give feedback, too;

- To accept feedback—The recipient should accept it as an input useful to consolidate or improve the relationship with the others. Accepting the feedback does not mean agreeing with it, but just being open to the perception of the reality that the others have. This is necessary for the individual to reframe his vision by changing its premises (Watzlawick, Weakland, and Fisch, 1974);
- To decline feedback—When the feedback is given in an inappropriate way, the recipient should decline it and give counter feedback to the provider to reorient the manner of presenting the feedback. The refusal can concern either the form of the feedback (e.g., no improvement sugges-

tions but just heading of mistakes) or the focus (e.g., misplaced positive or improvement feedback on being). Refusing the feedback does not deny the usefulness of the other's point of view, but tries to ensure that the other's perception is expressed in a way that is acceptable for the recipient.

3.2.4 Decision-Making and Strong Consensus Protocol

The efficiency, effectiveness, and success of a change process depend on the capability of the driver and his team in making sound decisions and carrying out the action plan with perseverance. Critical decisions are: who initiates the various forms of decisions, who makes the decisions, who confirms them, and who supervises their implementation. The contextual nature of the M3C requires the decision-making process to be consistent with the features of the specific context. Regardless of the way the decision is made, in the implementation phase, each member must commit to keeping the efforts aligned and to fully implementing the action plan. From a pragmatic point of view, the decision-making process can be separated into three steps (Table 3.2.4), as detailed in the next table:

- Phase one, divergence (to gather information and create options);
- Phase two, convergence (to screen solutions and choose the most appropriate);
- Phase three, commitment (to create consensus to design and implement the action plan).

In the divergence and the convergence phases, the driver can rely on analytical tools and methodologies (such as the real options approach) to optimize the efficiency and the effectiveness of strategic thinking. However, the driver should always remember that the power structure of the firm heavily affects the process of strategy formation and decision-making (Valverde, 2004). The divergence phase and the convergence phase can be carried out by few or many people. Crossing these two dimensions generates a taxonomy of four steps (phases 1.2 to 2.2) of the decision-making protocol described above. Such taxonomy applies to all the three basic hierarchical levels of a firm:

- At the corporate level, it concerns the process of defining the overall mission of the company and the allocation of the resources;
- At the business level, it makes reference to the design of the activities necessary to enhance business units' competitive position;
- At the functional level, it impacts the process of creation, development, and deployment of the competencies needed to sustain competitive advantage.

Each type of decision-making process has its advantages and drawbacks, depending on the context within which people operate. Table 3.2.5 illustrates the taxonomy for the corporate level with regard to the mission definition and resource allocation processes:

Divergence	Phase 1.1 Information Gathering	*What matters is collecting as much information as possible*, regardless if they come from inside or outside, top or bottom of the organization. Everybody is entitled to provide information, unless the specific feature of the context requires a restriction to participation. Looking at one issue from several points of view helps people to step back and to broaden their perspectives. It allows a great deal of participation, which contributes to support people's commitment in the implementation phase. In the case of an extremely urgent and important issue, the degree of participation can be reduced, but this must be made explicit. *No decision is made yet at this stage.*
	Phase 1.2 Options Creation	*What matters is to produce as many options as possible.* Options are ranked by making reference to their *feasibility*, with little concern for the degree at which they contribute to protect the stake. Exploring possibilities and keeping it open before making any decision is important because if something is feasible, it does not mean that it is the appropriate solution. Such an approach helps to not commit too early to an action that can reduce the flexibility of the firm in dealing with future changes in its external environment or internal context, and therefore to cope with uncertainty (Copeland and Keenan, 1998). To calculate the value of the options, Copeland and Tufano (2004) propose to use a simple and effective tool as the binomial model designed by Cox, Ross, and Rubinstein (1979). *No decision is made yet at this stage.*
Convergence	Phase 2.1 Solution Finding	*What matters is to transform an option into a solution.* Options are screened and ranked by making reference to the degree at which they *help to protect the stake*. In this phase participation should be restricted to the people that are directly responsible for the stake. It is important to look at the proposed options independently from their feasibility. In fact, it might happen that something that it is not feasible (e.g., because of a lack of resources), could be what is necessary to protect the stake. *Still, at this stage no decision is made yet.*

	Phase 2.2 Choice	*What matters is that people who are directly responsible for the Stake make the decision.* This means that they identify the appropriate set of actions necessary to protect the stake by comparing the content and outcomes of the options and the solutions proposed and by taking into account the ranking of what is feasible with the ranking of what is most likely to protect the stake. Then they explicitly announce which solution has been chosen, which resources are going to be allocated, what the milestones of the action plan are, the indicators chosen to check the performance, and the control and feedback procedures put in place to keep the process on track.
Implementation	Phase 3.1 Action Plan Design	*What matters is to make sure that everybody will act accordingly to the decision made.* In fact, many decisions do not provide the expected results because of flaws in their implementation (Valverde, 2004). Steps 1.1, 1.2, and 2.1 make as many people as possible contributing to the decision-making process. This phase is then necessary to allow people to take ownership of the project by letting them design the action plan that they will have to implement.
	Phase 3.2 Commitment Agreement	*What matters is that agents act accordingly to the decision made.* Every individual concerned by the implementation of the action plan has to commit to act consistently and up to the complete implementation of the action plan. The protocol to achieve such strong consensus is described by a decision tree in a later page.

Table 3.2.4—The decision-making protocol

		Convergence	
		Few	Many
Divergence	Many	*Participative Oligarchy* The definition of the mission is the result of a participative process where the employees at the business level and functional level are required to provide suggestions. The screening of the suggestions, the decision and the allocation of resources are made by the senior executives. *The budgeting process is often an example of a participative oligarchy decision-making approach.*	*Cooperative Democracy* The definition of the mission, the screening of the suggestions, the decision and the allocation of resources are the result of a participative process where the employees at all levels (corporate, business, and functional) are asked to provide suggestions and to take ownership of the process of strategy formation and implementation. *The labor contract negotiating process between trade unions and the firm are often an example of a cooperative democracy decision-making approach.*
	Few	*Technocratic Oligarchy* The definition of the mission, the screening of the suggestions, the decision, and the allocation of resources are made by few specialized people, and then enforced by executives by using the power and authority structure of the company. *A merger and acquisition process driven by a specialized bank or consulting agency is often an example of a technocratic oligarchy decision-making approach.*	*Technocratic Democracy* The definition of the mission is the result of a process where specialized people provide suggestions based on their expertise and knowledge of the specific context of the firm and its environment. The screening of the suggestions, the decision, and the allocation of resources are made then by people at all levels (corporate, business, and functional). *The management of a layoff process is often an example (in France) of a technocratic democracy decision-making approach.*

Table 3.2.5—The features of a decision-making process

The driver seeks to create the conditions, which allow as many people as possible to participate in the decision-making process (Choo and Bontis, 2002a), recognizing that a participative process may not be the most efficient and effective way to make a decision because:

- Even if participation implies that decision-making is shared, control is no less present, albeit in more subtle forms which could go against the identity of employees (Ezzamel and Willmott, 1998);

- Some contexts require decisions to be made quickly or to be kept confidential, which does not fit with the features of a participative oligarchy decision-making process.

Cooperative democracy is a more sophisticated form of the participative oligarchy decision-making process. In cooperative decision-making processes, the authority for major decisions is vested in all the participants, and the mode of internal decision is essentially democratic.

The mode of final commitment that best fits the constructs of cooperation, interdependence, and mutual reciprocity is consensus (Schutz, 1958, 1966, 1982, 1984). In fact, organizations are social communities that are capable of coordinating human behaviors better than market contracts because they provide "voluntaristic [sic] actions structured by organizing principles that are not reducible to individuals" (Kogut and Zander, 1992, p.384). From this perspective, consensual commitment provides the community with more creative solutions, even if the process may last longer. Consensual action plans are usually executed more efficiently and effectively, since all members who will implement the decision have participated in the process.

A consensual commitment is made when all members agree with the action plan even if some disagree with the decision made on the issue. Consensus differs from unanimity (all members agree on the action plan), from majority (more than half of the members agree on the action plan), and also from authoritarianism (just one person agrees on the action plan).

The strong consensus protocol (Diagram 3.2.2), is useful for carrying out the commitment phase under the construct of cooperative democracy.

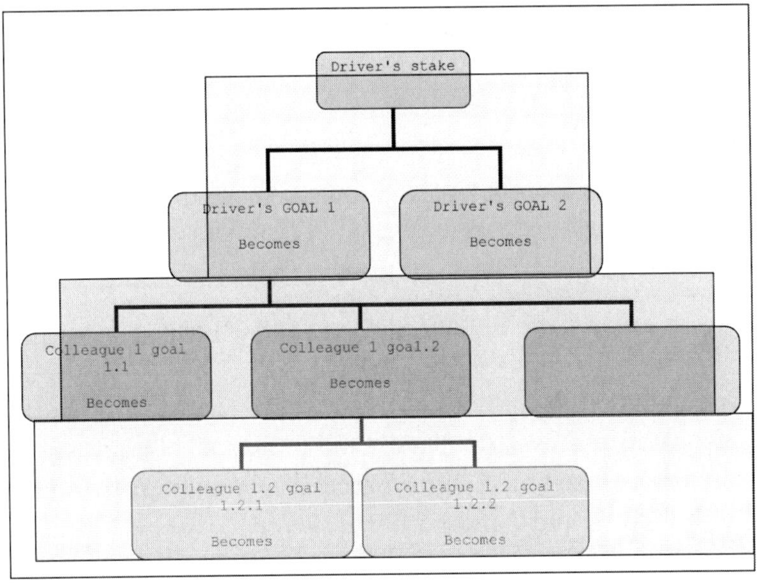

Diagram 3.2.2—Strong consensus protocol

3.3 CONTEXTUAL DRIVERSHIP

A dynamic, uncertain environmental context calls into question the appropriateness of the traditional "and/or logic" hierarchy regarding leadership and management roles. The M3C model offers drivership, which melds leading and managing as an effective modality for governance within a dynamic context. In the drivership paradigm, the driver's role is both symbolic and operational in regard to establishing and sustaining stability and the quality of the relationship with colleagues. The driver and colleagues assume the responsibility for establishing relationships based on an acceptable level of autonomy. The driver provides the appropriate guidance determined by the assigned level of autonomy, regardless of the hierarchical position of the colleague.

The drivership paradigm provides organizations with a coherent governance process, which offers a sense of stability during the experiencing of change. drivership provides the individual with an emotional anchor by clearly specifying the degree and level of responsibility. Drivership answers the question of "What am I responsible for in regard to the stake as the organization evolves?"

Appendix two shows the interdependence of all the contextual drivership protocols described in this chapter.

3.3.1 Organizational Alignment Map

Defining a meaningful vision to be realized and gearing it down into sound and consistent team and individual objectives is one key for organizational performance and business success (Bass, 1985; Kaplan and Norton, 1992, 1996; Conger and Kanungo, 1998; Berson, Shamir, Avolio, Popper, 2002). Shaping a vision means asking people to take a risk by changing the actual way of doing business into a new one, which can fit better to the forecasted future environment and context (Prahlad and Hamel, 1994a, 1994b). For the vision to be realized, the associated stake identified by the driver must be accepted and internalized by all people concerned by the change process (Fischer, Ury and Patton, 1981, 1991; Lax and Sebenius, 1996).

Individuals, teams, organizations, and stakeholders' interests should be prioritized and coordinated to ensure the consistency of the actions aligned and translated into consistent objectives (Hersey and Blanchard, 1974; Kaplan and Norton, 1992, 1996; Wilderom, Glunk and Maslowski, 2000; Freeman, 1984; Tsui, 1990). The organizational alignment map and the driver's added value matrix aim to ensure consistency at all levels of the organization of the actions taken to make the vision becoming reality. These protocols confront and merge the principled negotiation (Fischer, Ury and Patton, 1981, 1991; Lax and Sebenius, 1996) and the management by objectives (Drucker, 1954; Hersey and Blanchard, 1974) approaches, as shown in the Table 3.3.1.

Principled Negotiation	Management by Objectives
Stake to be protected	*Vision to be realized*
Interests to be satisfied	*Goals to be met*

Table 3.3.1—Contrasting principled negotiation and management by objectives approaches

Setting alignment goals is insufficient to mobilize people's commitment and competencies: the MBO approach is interesting only if the dynamics of goal-setting are correct (Blake and Mouton, 1985; Wiley and Brooks, 2000). The process of transferring responsibilities from the driver to the colleagues takes the form of a contract negotiation (Hersey and Blanchard, 1974; Lax and Sebenius, 1996) tailored to an individual's and/or team's reliability level and objectives. Under the contextual drivership protocol, the contract is open for renegotiation by either party when the context changes or if goals are not met. The appropriate setting of the organizational alignment map requires clarifying the differences in the meaning of the terms stake, goal, result, and purpose, and their reciprocal relationships in terms of priority and impact on individuals,' teams' and organizations' work.

Identifying the stake to be protected is necessary to shape a meaningful vision and draft a sound organizational alignment map. As defined in part one, the stake is the most valuable tangible or intangible asset that an individual or a community wants to protect with their action. While interests can be traded off, the stake cannot. The stake is the ultimate non-negotiable interest (Fischer, Ury and Patton, 1981, 1991). If the context changes it may happen that the stake changes (and vice versa). The stake is the fixed reference that defines the direction and orientation of all actions, which contribute to the stake. Clearly identifying the stake helps drivers build and communicate the vision to be realized. To ensure the consistency of individual and collective efforts, it is important to have only one complementary stake for each level of decision and for each field of action. Focusing on the stake helps avoid paralysis in decision-making processes and optimizes the leverage of resources (Collins, 2001).

Articulating the goals to be achieved is the second step in building a sound organizational alignment map. Systemic theory argues that goal-setting can effectively drive the action of an individual in the long run, only if achieving the goals depends on this individual (de Saint Paul and Tenenbaum, 2002). This implies that:

- The objective has to be chosen by the individual;
- He must have the power and the means to take the actions necessary to achieve the goal;
- The outcome of the goal must be ecological (both the individual and the other interdependent actors in this context are better off when the individual's goal is achieved).

A goal must be translated into concrete actions, which satisfy the prioritized interests and can be described in terms of scope and span. (Blake and Mouton, 1985; Kaplan and Norton, 1992, 1996), as shown by Table 3.3.2. Four types of goals are used in the autonomy curve to adapt the drivership style of the driver to the reliability of the colleague.

<table>
<tr><td>Scope and span</td><td>Small</td><td></td><td></td><td>Large</td></tr>
<tr><td rowspan="13">Type of goal</td><td>Instructions</td><td>Steps
(set of instructions)</td><td>Tasks
(set of steps)</td><td>Missions
(set of tasks)</td></tr>
<tr><td>Instruction 1</td><td rowspan="2">Step 1</td><td rowspan="6">Task 1</td><td rowspan="12">Mission 1</td></tr>
<tr><td>Instruction 2</td></tr>
<tr><td></td><td rowspan="4">Step 2</td></tr>
<tr><td></td></tr>
<tr><td></td></tr>
<tr><td></td></tr>
<tr><td></td><td rowspan="3">Step 3</td><td rowspan="6">Task 2</td></tr>
<tr><td></td></tr>
<tr><td></td></tr>
<tr><td></td><td rowspan="3">Step 4</td></tr>
<tr><td></td></tr>
<tr><td>Instruction n</td></tr>
</table>

Table 3.3.2—Type of goals to fit the drivership protocols

Teamwork is needed to implement a strategic decision. Therefore, goals that are competitive or individualistic should be avoided. Individualistic goal structures offer little or no relationship to members' goal achievement. Competitive goal structures generate negative relationships. Competitiveness provokes fear of failure, low trust, and win-loss strategies. In a cooperative goal process, the relationship among each member's goal achievement is positive. Cooperation leads to awareness of mutual interdependence, effective communication, and exchange of information, which facilitates each other's productivity. Setting cooperative goal structures must be founded on a conception of leadership as a service (Greenleaf, 1977; Johnson et al, 1981; Johnson and Johnson, 2000; Isaken and Lauer, 2002).

Results or outcomes happen because of the actions taken in a given context and measure the level at which goals are met.

To provide the colleague with a global, meaningful framework without constraining his freedom in implementing the action plan, the driver should be able to make explicit one or several purposes to be pursued for each identified goal. At any decision-making or operational level, defining the stake and the purposes corresponds to a strategic decision while setting goals corresponds to a tactical decision. A purpose is always described by verbs of action (deSaint Paul and Tenenbaum, 2002).

From a practical point of view, to establish the organizational alignment map, the driver and his colleagues must:

1. Identify the stake and translate it into a consistent vision.

2. List and prioritize the interests and gear them into several operational goals to be met by each colleague. To provide a global framework for meaningful action, the purposes of each goal must be made explicit and the criteria used to measure its outcome must be identified.

3. Each goal becomes a stake for one or several colleagues who must act as drivers and repeat steps one and two for their colleague, as shown by Diagram 3.3.1.

This cascading of goal responsibility can be applied to the management of project teams, where the hierarchical relationship is usually not present, or is bypassed.

An interesting example of the difference between the stake and the goal is provided by Alstom's victory in the bid for the Korean high-speed train in the mid '90s. Alstom's president and CEO said that the bid was a success for Alstom because the stake was safe, even if the financial and economical results of the negotiation where far from what was forecasted.

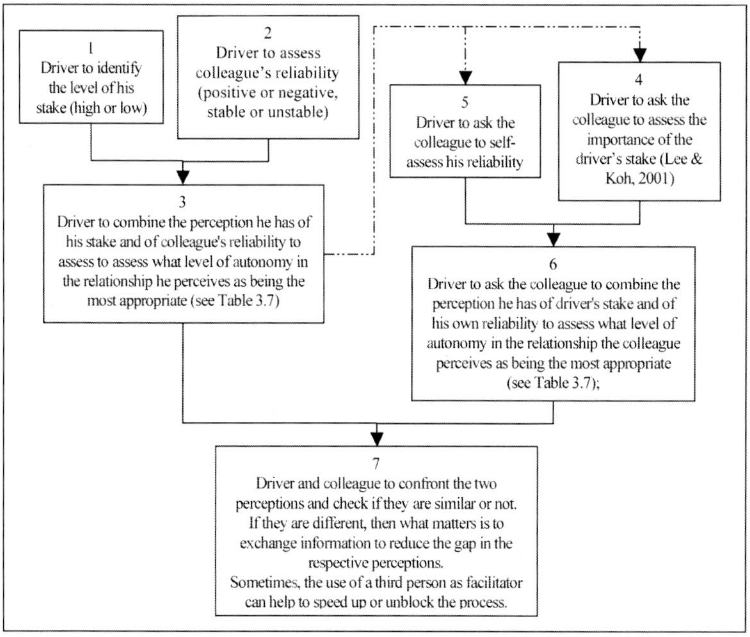

Diagram 3.3.1—Goals' cascade to design the organizational alignment map

Alstom's stake was to establish TGV technical features as the standard in the Far East Asia fast-growing rail transportation market, to position itself for future bids in China, Thailand, and India.

Korean TGV bid's purpose was mainly to settle commercial/manufacturing operations in a highly industrialized country within the region and create the production network necessary to provide high-quality products at reasonable costs.

Korean TGV bid's goals were multiple, the most important being profitability and maximization of products manufactured in French factories. However, Alstom chose to trade off these interests to protect its stake. Indeed, the outcomes of the negotiation have been: zero expected profit (but the money recovered from contingencies); only one million working hours over the expected three million allocated to French plants, with the two other million going to Korean subcontractors. In this way Alstom has been able to:

- Win the bid against its competitor Siemens;
- Train Korean subcontractors to TGV technology;
- Insert them in the vendor lists for future high-speed railways bid in the emerging countries of the region.

3.3.2 Driver's Added Value Matrix

To increase the performance of the organization, the organizational alignment map helps drivers separate, at each operational level from the board to the front line operators and employees, the strategic dimension, the stake to be protected, and the vision to be realized from the tactic dimension, the interests to be satisfied, and the goals to be met.

Performance is a function of efficiency and effectiveness. Efficiency is an internal measurement concerning the relationship between organizational input and organizational output. It assesses the way resources—organizational, human, financial, technological, and material—are used to meet goals related to specific strategies. Efficiency is an indicator of the health of an organization and its capability of optimizing the allocation of available resources to obtain specific results. Effectiveness is an external measure concerned with the capability to get the results defined by the strategy and rewarded by the external environment. Effectiveness assesses outcomes, not the way they are obtained (Pfeffer and Salancik, 1978; Porter, 1996).

These two dimensions, although related, are different in nature, with no straight link between them. Regardless of how efficiently a company uses its resources, it might produce products that are not required by the market. An increase in efficiency will not increase the effectiveness of its product in the market. The opposite can also be true. A company might have a product that sells but with high production costs, which results in financial losses.

The driver's added value matrix (see Table 3.3.3) maps the complexity of the activities' interfaces and the interdependence of the goals. It attempts to optimize the performance of all the system's actors by clearly identifying which actions have to be carried out by whom. Used in connection with the balanced scorecard approach (Kaplan and Norton, 1992, 1996), it helps to keep the individuals aligned in their action to protect the stake and realize the vision. The driver's added value matrix is the combination of two action dimensions: execution and planning. The execution dimension takes into account the "do/make do dilemma." The planning dimension takes into account the "certainty/uncertainty dilemma."

Over-planning diminishes the adaptability of the organization to exogenous change. Too little planning leads to confusion in task assignment and failures in interface coordination. Leveraging organizational factors (team structure, project planning) helps to move the certainty/uncertainty line (Prahlad and Hamel, 1994a, 1994b).

If the driver is involved in too many actions, his capability to focus on what is important to protect the stake is reduced. Too few actions reduce the driver's capability to control others' actions and results. Leveraging drivership protocols, especially the reliability matrix and autonomy curve, contributes to move the do/make do line (Hersey, 1984; Hersey and Blanchard, 1969, 1972, 1974; Blake and Mouton, 1985).

		Actions' Planning	
		Uncertain	Certain
Actions' Execution	Do	*Crisis Activities* All those actions that: - Cannot be planned because of their high level of uncertainty (or because they are unknown); - Can be carried out only by the driver because of their big impact on the stake. The driver is directly in charge of crisis activities, or he has to drive his colleagues to face them by acting according to the most suitable scenarios activities already prepared.	*Core Activities* Set of actions that must be done by the driver and that can be planned with a reasonable level of certainty, like: - To decide/plan the appropriate actions to protect the stake given to him; - To identify the subsidiary activities by making reference to the interests to be satisfied and the goals to be achieved; - To use the reliability matrix and the autonomy curve to identify the right colleagues to execute each subsidiary activity and the way of supervising them; - To assess colleagues' performance and decide consequent sanctions (rewards or penalties); - To make sure that colleagues set up suitable scenarios; - To manage crisis activities.

		Scenarios Activities	*Subsidiary Activities*
	Make Doing	Aimed to reduce reaction time facing a crisis by identifying possible patterns of actions that can affect the stake, and that cannot be planned because of their level of uncertainty. A scenario describes a possible future state of the reality and why/how that state might develop. It is a hypothetical sequence of events constructed for the purpose of focusing attention on causal processes and decision points. It is neither a prediction nor a belief of what the future will actually look like; it helps people think about potential threats and opportunities differently than they have the habit to do. (Schoemaker, 1995; Smith, 2003). Allocating resources to scenarios sactivities is one of the most important and the most disregarded responsibilities of a driver (Prahlad and Hamel, 1994). To increase the probability of successful implementation, scenarios activities must be designed by the colleagues that will have to execute them when the issue arises. The driver has then to: - Identify the uncertain events that could impact the stake and ask for the setting of one or several scenarios; - Drive the colleagues to set the scenarios process (design, decision-making and implementation) and to check its consistency and applicability; - Order and follow-up the implementation of a scenario activities when a crisis arises.	All foreseeable activities that can be carried out by colleagues and that the driver controls. The driver should not design the subsidiary activities on the basis of what he does not like to do, but on the basis of who can bring the higher added value in executing such activities (Hersey and Blanchard, 1974). Under the drivership paradigm, it may happen that a driver's subsidiary activities have to be carried out by his hierarchical supervisor, if the organizational structure of the company requires so. The typical example is the budgeting process resource allocation, where the subordinate (driver) needs his supervisor to give him the resources before being able to implement the actions he is assigned to do to protect the stake.

Table 3.3.3—Driver's added value matrix

Arbitrating the priorities between urgent activities and important activities is a critical driver's responsibility, as shown by the time management matrix (Table 3.3.4):

		Urgency of the task (impact on the workload)	
		High	Low
Importance of the task (impact on the Stake)	High	To be done first	?
	Low	?	To be done last

Table 3.3.4—Time management matrix

The driver adds value when there is balance between the two dimensions—importance/urgency and/or execution/planning:

- The biggest possible subsidiary activities cell; that is, a lot of not urgent/not important (for his level of responsibility) activities carried out by his colleagues (for their level of responsibility the same activities are not urgent/important)
- The smallest crisis events cell (the fewest urgent/important activities to be carried out by himself);
- A core activities cell as horizontal as possible (important activities to be carried out without urgency).

Urgent/important activities must be carried out first; not urgent/not important activities can be carried out last. With regard the two other cells, should the driver give priority to urgent/not important activities or to not urgent/important ones? In Table 3.3.5 the time management matrix (in italics) is coupled with the driver's added value matrix to facilitate the answer to such questions. When the added value matrixes of the driver and of his colleagues are properly set up, the core activities (low urgency/high importance) are carried out just after those in the crisis events cell. This allows the driver to pilot the implementation (by his colleagues) of the most suitable scenario to face urgent/not important actions.

		Actions' planning *Urgency of the task* *(impact on the workload)*	
		Uncertain *High*	Certain *Low*
Actions' execution *Importance of the task* *(impact on the stake)*	Do *High*	Crisis activities *To be done first*	Core activities *To be done second*
	Make/do *Low*	Scenarios activities *To be done third*	Subsidiary activities *To be done last*

Table 3.3.5—Coupling the time management matrix and the driver's added value matrix

3.3.3 Reliability Matrix

"Learning does not occur in any enduring fashion unless it is sparkled by people's own ardent interest and curiosity. When the spark is not present, people compliantly accept training in a subject … The effects of that training last for a while, but without commitment, the trainees stop using the new skills" (Senge et al., 1994, p.193). The constructs of motivation/commitment and competencies/ability have generated discussion in the management literature in regard to the characteristics of an individual, which may be related to performance (Hersey and Blanchard, 1972, 1974, Tissier, 1988, Vroom, 1964; Hersey, 1984).

Commitment is the individual's state of being, which is grounded on his awareness of the context. No matter what the satisfiers and dissatisfiers factors present in the context, it leads the individual to explicitly and voluntary leverage his own physical, intellectual, emotional, and intuitive resources to protect his own and his team's stakes and to contribute to his own and his team's actions and accomplishments (Herzberg, 1958, 1966; Becker, 1960; Buchanan, 1974; Porter, Steers, Mowday and Boulian, 1974; Salancik, 1977a, 1977b; Wiener, 1982; Meyer and Allen, 1997; Kanter, 1968).

Competence is the balanced and appropriate combination of knowledge, reflective experience (know how), and behaviors that an individual uses to carry out a given activity in a specific context (Nonaka and Takeuchi, 1995, 1996; Senge et al., 1994).

Commitment enhancement and competencies development are two interdependent and dynamic concepts. In the contextual drivership framework, reliability is the assessed capability of an individual to persistently demonstrate commitment to the stake and to leverage the competencies appropriate to the activities required within the context. The level of reliability is directly related to the commitment and the competencies of the colleague, as shown by the following formula:

Reliability = f (commitment of the individual, competencies of the individual)

The stability of reliability describes the likelihood for the colleague to persist in his logic of functioning for a reasonably long span of time.

Combining the commitment and competencies dimensions into the reliability matrix facilitates the deepening of the analysis of the stability of the colleague's reliability. Assessing reliability is mainly a qualitative process, based largely on the driver's capability, through appropriate communication, to detect a colleague's commitment and competency to carry out the various activities.

The combination "high commitment/high competencies" leads to a stable virtuous circle of continuously-increasing positive reliability.

"Low commitment/low competencies" leads to a self-reinforcing vicious cycle towards a stable and continuously decreasing negative reliability (Hersey and Blanchard, 1972; Hersey, 1984).

The combinations "high commitment/low competencies" or "low commitment/high competence" result in an uncertain level of reliability, which depends on the difficulty of the task and the interests of the colleague in the activity's outcome. Reliability is unstable. Its evolution depends on the effects that changes in the context and activities have on both commitment and competence. All systems tend toward expending the lowest possible level of energy. Therefore, entropy increases if the importation of external energy is less than that used. If action is not taken to support Reliability, progressive worsening of commitment and competencies becomes an issue. Reliability diminishes over time within the southwest cell, resulting in dependence. Table 3.3.6 presents the four cells of the reliability Matrix, which are labeled by making reference to:

- The features of the reliability (stable or unstable, positive or negative);
- Four logics of functioning of the colleague with others, which describe his approach to professional relationship: interdependence, dependence, pseudo-dependence, independence;
- The four ways of relating to people, which they describe how the colleague conceives interpersonal relationships: we together, you for me, you with me, for myself alone (Barjou, 1999).

Assessing the level of reliability of the colleague is a necessary preliminary step to identifying the level of autonomy of the relationship between the driver and the colleague. Therefore, the drivership strategy adapts to fit to the context, to the activity, and to the individual. A driver must have a holistic approach in regard to a colleague and adapt his drivership style to his colleague's reliability in each of the activities to be carried out. In fact, the same colleague can show different levels of reliability because of differences in the context and the activities. The systemic feature of reliability implies that all levels are interdependent: the improvement/worsening of reliability in one activity could imply the improvement/worsening of reliability in some others.

level of Colleague's commitment	High	The combination of a high level of commitment and a low level of competencies results in an *Unstable Uncertain Reliability* Inversely related to the difficulty of the activity compared to colleague's competencies. When an activity does not requires specific competencies, his high level of commitment helps him to overcome difficulties without needing help from others. In this case, his level of reliability has a *positive trend*. When the activity requires the use of specific competencies that he cannot leverage, his willingness to do things is not enough to carry out the activity without the help of others. In this case, his reliability has a *negative trend*. The colleague would like to be able to perform the activity without asking for support, but cannot do much without having somebody to help him. His logic of functioning with others is: *Pseudodependence* The way he conceives professional and human relationships with people is: *"You with me"* Colleague's specific observable behaviors are: - He shows interest and asks for help to improve; - He needs instructions on what to do and how to do it; - He has to be reassured on his contribution/role.	The combination of a high level of commitment and a high level of competencies results in a *Stable Positive Reliability* directly related to colleague's perseverance in and capability of leveraging appropriate competencies. He understands the systemic nature of performance and is aware that it depends as well on others' contributions (Lax and Sebenius, 1986). His logic of functioning with others is : *Interdependence* The way he conceives interpersonal relationships with people is: "*We together*" Colleague's specific observable behaviors are: - He is enthusiastic about the activity and loyal in relationships with others; - He is self-confident in goal achievement and knows how to make people feeling self-confident; - He has and leverages all needed competencies to perform the activity; - If he lacks some competencies, he knows how to seek them and/or leverage others' resources.

	Low	The combination of a low level of commitment and a low level of competencies results in an *Stable Negative Reliability* Directly related to colleague's lack of perseverance in and low capability of leveraging appropriate competencies. He does not understand the systemic nature of performance and is not aware of how his behaviors impact others' contributions. His logic of functioning with others is: *Dependence* The way he conceives interpersonal relationships with people is: "*You for me*" Colleague's specific observable behaviors are: - He is not interested by the activity and shows inappropriate behaviors with people; - He does not have and/or use the appropriate competencies to perform the activity; - He avoids responsibilities and proactive behaviors; - He is focused more on difficulties than on opportunities and solutions.	The combination of a low level of commitment and a high level of competencies results in an *Unstable Uncertain Reliability* Directly related to the interest the colleague has in the outcome of the activity. When he is interested by the activity's outcome, he will be keen to leverage all his competencies to carry out the activity. In this case, his reliability has a *positive trend.* When the activity outcome has no interest for him, his low level of commitment will push him to orient his competencies and energies to do things more related to his interests. In this case, reliability has a *negative trend.* His high level of competencies makes him able to successfully carry out the activity; his low level of commitment becomes an obstacle to his contribution to team's actions and performance. His logic of functioning with others is: *Independence* The way he conceives interpersonal relationships with people is: "*For myself alone*" Colleague's specific observable behaviors are: - He is self-confident, looks for freedom in decision-making and execution of activities; - He is stuck on his ideas, and has difficulties in listening and integrating others' contributions; - His attitude to work is highly dependent on mood and relationships with others.

Table 3.3.6—Reliability matrix

Team's Reliability Matrix

The concept of reliability can be used to drive teams. Team reliability is the feature of a group of people who are persistently capable of:

- Showing consistent cohesiveness in working together (despite possible divergences in individual interests), which means being able of concentrating their efforts on the protection of a common stake through coordinate, efficient, and effective actions.

- Creating synergies amongst all available and appropriate competencies to successfully carry out a given activity by taking into account the features of the context.

A high level of cohesiveness shows that team members are able to:

- Set up and abide to simple and co-opted ground rules of functioning and their corresponding positive and negative sanctions;
- Make explicit their own interests and the way they perceive problems (Fisher, Ury and Patton, 1981, 1991);
- Integrate others' perceptions to accept diversity as a resource (Senge et al., 1994);
- Focus on solutions instead of problems and build on others' proposals;
- Make explicit any disagreement and propose options and solutions acceptable for self and others (Fisher, Ury, and Patton, 1981, 1991);
- Once agreed on an action, commit individually and look for team's strong consensus (Schutz, 1958, 1966, 1982, 1984).

This set of behaviors ensures that each individual will push in the same direction, act to keep high the moral of people, and provide assistance to others.

A high "synergistic" team is able to create synergies and respect the expertise and contribution of each team member, and it is characterized by:

- A balanced set of competencies covering the entire scope of the activity or mission to be carried out; if all needed competencies are not available in the team's members, external competencies will be sought (Nonaka and Takeuchi, 1995);

- A clear definition of the interfaces and of the way of making different competencies cooperating (Senge et al., 1994; Grant, 2001);

- An equilibrium between individual expertise/specialization and multi-skill versatility (Nonaka and Takeuchi, 1995);

- The capability of integrating new competencies and reshaping old ones, eventually by looking outside of the individuals' and team's field of expertise and business (Senge et al., 1994; Nonaka and Takeuchi, 1995; Aubrey, 1994).

This set of behaviors ensures that the team will approach issues and actions from multiple points of view and will be able to use the most suitable resources to achieve its goal in an efficient and effective way.

Implementing a Drivership Strategy to Effectively Develop Reliability

Reliability is not a fixed feature of a colleague. Using the reliability matrix does not mean putting people in boxes or encouraging the driver to adopt a rigid and thoughtless approach to a colleague's assessment. The driver has to understand and accept the fact that the reliability of a colleague can be variable.

Assessing the reliability of a colleague should not be aimed to judge the person, but to facilitate and foster the sharing of own perceptions between the driver and the colleague (Porter and Lawler, 1968; Lax and Sebenius, 1986). When discussing reliability, people should make reference to facts (Hersey, 1984).

Commitment must be assessed by making reference to its features (visibility, irrevocability, and volition) that can be translated into observable behaviors (Salancik, 1977a, 1977b). The three dimensions of competence (knowledge, reflective experience/know-how, and behaviors) must be assessed by making reference to factual elements.

The driver has to leverage all the factors of reliability (knowledge, reflective experience, behaviors, comfort, and stimulation) to create and maintain positive evolutionary dynamics toward the northeast cell and increase individual and team reliability. At the same time, he should make sure that the existing level of commitment and competencies does not decrease to avoid the fall of the reliability toward the southwest cell.

Boosting individual and team reliability is necessary to allow the establishment of driver-colleague relationships characterized by an appropriate level of autonomy. This process allows the driver to expand the colleague's activities portfolio (Nonaka and Takeuchi, 1995) and refocus core activities in regard to what is important and not urgent.

Table 3.3.7 shows what should be, for each cell, the purpose of the reliability development strategy put in place by the driver and fitting to the type of functioning of the colleague:

Individual and team's reliabilities need not be correlated. It may happen that in a high-reliability team, some of the members have low reliability, or on the contrary, that in a low-reliability team, some individuals show high reliability. The driver must take it into account and adopt different approaches to drive the team as a whole while customizing his drivership to each colleague, as described later by the autonomy curve protocol.

		Level of competencies	
		Low	High
Level of commitment	High	*Unstable, uncertain level of reliability* *Recognize and reward commitment by:* nurture the comfort factors that fuel colleague's commitment praise the colleague for the effort done and avoid stressing him with short-term performance request *Create and develop competencies by:* provide the colleague with factual feedback on his performance and explain what should be done differently make explicit to him the corrective actions that must be taken to improve performance and get the required outputs	*Stable, high level of reliability* *Encourage and reward commitment by:* nurture the comfort factors that fuel colleague's commitment provide the individual with appropriate activities that leverage stimulation factors *Consolidate, leverage, and spread competencies by:* recognize colleague's knowledge and help him to transfer it to others leverage colleague's skills, make him formalize them and allow him to make new experiences emphasize the exemplarity of colleague's virtuous behaviors
	Low	*Stable, low level of Reliability* *Create condition for commitment by:* make sure the comfort factors are in place listen and leverage colleague stimulation factors bind the Colleague to abide to co-opted rules and related sanctions; *Create basic competencies by:* provide the colleague with appropriate training to acquire the basic knowledge create occasions to help the colleague develop operational skills and build his know-how provide reinforcement feedback for expected behaviors and reorienting feedback for unwanted ones	*Unstable, uncertain level of reliability* *Facilitate and enhance commitment by:* improve as much as possible the comfort dimension listen and leverage colleague's stimulation factors bind the colleague to abide to co-opted rules and related sanctions *Recognize and leverage competencies by:* push others to solicit colleague's knowledge and encourage him to transfer it to others explore colleague's skills and ask him to propose ways to adapt it to actual needs provide reinforcement feedback for expected behaviors and reorienting feedback for unwanted ones

Table 3.3.7 - Strategies to improve reliability

3.3.4 Autonomy Curve

Assessing the reliability of a colleague is not enough to decide what drivership style has to be adopted. Contrary to what was proposed by Hersey and Blanchard (1972, 1974), the fact of facing a reliable colleague should not automatically lead the driver to transfer him responsibilities. Colleague's reliability is only one of the factors influencing the level of autonomy in the relationship, the other being the driver's stake. The drivership paradigm uses the construct "level of autonomy" (from now on autonomy) to determine the type of relationship. In a given context, the driver (principal) and the colleague (agent) mutually choose to establish to properly initiate and regulate the actions necessary to protect the driver's stake by leveraging the colleague's reliability.

The autonomy curve protocol ensures the necessary systemic links between the needs of the driver in protecting his stake and the resources of the colleague in carrying out the required actions. In addition, it can be used to assess the overall functioning of a team in implementing a task, an action, or a project. By linking the features of the colleague (commitment and competencies) and the stake of the driver, the construct of autonomy overcomes the major limitation identified in the situational leadership approach while it provides an operating tool to foster accountability.

Autonomy is directly related to the reliability of the colleague, and inversely related to the stake of the driver, as shown by the following formula:

Autonomy = f (Reliability of the colleague, 1/Stake of the driver).

Identifying the appropriate level of autonomy means linking the stake of the driver and the reliability of the colleague, as shown by Table 3.3.8. The perception the driver has of his own stake is biased by the perception he has of the reliability of the colleague, and vice versa. If the driver grants to the colleague positive reliability, a reduction factor affects the perception the driver has of the degree and extent of control to exert over the colleague. If the driver grants to the colleague negative reliability, the perceived need to control increases.

		Colleague's Reliability			
		Stable negative	Unstable negative trend	Unstable positive trend	Stable positive
Driver's Stake	High	Low Autonomy	Limited Autonomy	Limited/Sufficient Autonomy	Sufficient Autonomy
	Low	Limited Autonomy	Limited/Sufficient Autonomy	Sufficient Autonomy	High Autonomy

Table 3.3.8—Relationship between driver's stake and colleague's reliability to identify the level of autonomy

Diagram 3.3.2 shows the process the driver and the colleague has to follow to assess the proper level of autonomy in their relationship. Steps one through three allow the driver to become aware of the type of relationship he would like to establish; steps four through six allow the colleagues to do so. Step seven allows the driver to confront his perception with his colleague's perception, so they can identify the most appropriate level of autonomy (Hersey and Blanchard, 1974, Porter and Lawler, 1968; Lax and Sebenius, 1986).

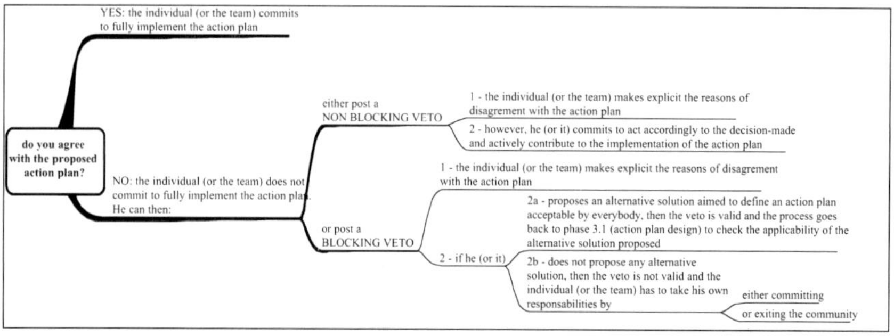

Diagram 3.3.2—Process to assess the level of autonomy in the relationship between the driver and the colleague

The next paragraphs are dedicated to explain how to use the autonomy curve protocol. Tables 3.3.9, 3.3.10, 3.3.11, 3.3.12, 3.3.13 show how the autonomy curve protocol facilitates the implementation of an MBO contractual approach (Hersey and Blanchard, 1974). Diagram 3.3.3 details its features and shows the interdependence of the parameters stake and reliability with features such as the drivership style; the role the colleague expects the driver to play; the colleague's main need in guidance; the MBO features: type of goals to be set and type of control to be carried out; the communication's features; the fixed points (FIP) follow-up and slots of active support (SAS) protocols. The drivership style represents the way the driver runs the relationship with the colleague, and it can be described by the role the driver has to play, the main colleague's guidance need, the type of goals, and the type of control the driver and the colleague agree on. In any relationship, each actor plays a role and expects the counterpart to play a complementary role, consistent with the context (Watzlawick, Beavin, and Jackson, 1967). In the drivership paradigm, the driver must be aware that playing the appropriate role has a high symbolic value for the stability and the soundness of the relationship with the colleague. Playing a role is not enough: the driver must provide the appropriate type of guidance required by the level of autonomy of the relationship, no matter what the hierarchical position of the colleague is. In a high autonomy relationship, if the driver plays the role of an instructor, most probably the colleague will feel constrained and frustrated, because his main need is orientation. In a low autonomy relationship, if the driver plays the role of a mentor, most probably the colleague will feel lost and anxious because his main need is direction. In both cases, the relationship will be unstable and unhealthy in the long term.

Level of Autonomy of the relationship		*Low*	*Limited*	*Sufficient*	*High*
Drivership style (to develop the colleague's reliability)		Teaching	Training	Delegating	Empowering
Role the colleague expects the driver to play		Instructor	Tutor	Coach	Mentor
Main colleague's guidance need		Direction	Support	Participation	Orientation
MBO features	Type of goals to be set	Instructions	Steps	Tasks	Missions
	Type of control to be carried out	Complete	Systematic	Focused	Random

Table 3.3.9—Features of driver-colleague relationship with regard to each level of autonomy

Level of Autonomy of the relationship		**Low**	**Limited**	**Sufficient**	**High**
Drivership style to develop Colleague's Reliability		Teaching	Training	Delegating	Empowering
Role the Colleague expects the Driver to play		Instructor	Tutor	Coach	Mentor
Main Colleague's guidance need		Direction	Support	Participation	Orientation
MBO features	Type of goals to be set	Instructions	Steps	Tasks	Missions
	Type of control to be carried out	Complete	Systematic	Focused	Random

Diagram 3.3.3—Global view of the factors impacting the autonomy curve protocol

Level of autonomy of the relationship		*Low*	A high stake activity has to be carried out by a colleague who has a low level of reliability, which is originated by one of his logics of functioning: dependence, pseudo-dependence facing a difficult activity, independence facing a not interesting activity.
Drivership style (aimed to develop colleague's reliability)		Teaching	*Helping* the colleague to learn new competencies. *Improving* his commitment by *stimulating* his interest in the activity
Role the colleague expects the driver to play		Instructor	*Firmly directing* the colleague and *teaching* him appropriate operational behaviors by *giving clear instructions* and *thoroughly controlling* his actions and their outcome.
Main colleague's guidance need		Direction	Because of the colleague's low level of reliability, the driver must exert a strong direction to avoid chaos and reduce uncertainty, sources of stress and instability. To learn without anxiety or to act despite his low commitment, the colleague needs to be told what to do and how to do it.
MBO features	Type of goals to be set	Instructions	In organizing the colleague's activity, the driver subdivides it into elementary instructions, simple enough to be carried out after a short and factual explanation (Blake and Mouton, 1985).
	Type of control to be carried out	Complete	To protect his own stake, the driver has to fully control the colleague's execution of the instructions.

Table 3.3.10—Features of the relationship between the driver and the colleague with regard to a low level of autonomy

Level of autonomy of the relationship		*Limited*	A low stake activity has to be carried out by a colleague who has a low level of reliability, which is originated by one of his logics of functioning: dependence, pseudo-dependence facing a difficult activity, independence facing a not interesting activity.
Drivership style (aimed to develop colleague's reliability)		Training	*Allowing* the colleague apply existing and new competencies *Reinforcing* his commitment by giving more freedom and *rewarding* success
Role the colleague expects the driver to play		Tutor	*Patiently supporting* the colleague and *training* him *to implement* appropriate operational behaviors by *providing steps to overcome* and *systematically controlling* his actions and their outcome.
Main colleague's guidance need		Support	Because of the low stake of the activity, even if the colleague's level of reliability is low, the colleague's level of stress is lower. The driver can give more freedom and be available to help the colleague to overcome the difficulties he could face.
MBO features	Type of goals to be set	Steps	In organizing the activity of the colleague, the driver subdivides it into *steps (sets of instructions),* simple enough to be carried out by without a continuous explanation of each instruction by the driver (Blake and Mouton, 1985).
	Type of control to be carried out	Systematic	To develop the colleague's competencies and increase his commitment, at the end of each step the driver has to systematically control its execution and provide the colleague with a global feedback.

Table 3.3.11—Features of the relationship between the driver and the colleague with regard to a limited level of autonomy

Level of autonomy of the relationship		*Sufficient*	A high stake activity has to be carried out by a colleague who has a high level of reliability, which is originated by one of his logics of functioning: interdependence, pseudo-dependence facing a simple activity, independence facing a challenging activity.
Drivership style (aimed to develop colleague's reliability)		Delegating	*Transferring* to the colleague the power on the means (human, financial, material…) while *keeping* the overall responsibility of the outcome.
Role the colleague expects the driver to play		Coach	*Accepting* the colleague participation in decision making and *delegating* to him the action implementation by providing *extended tasks* and focusing the control on the *key factors of success* of the activity.
Main colleague's guidance need		Participation	Even if the stake is high, the colleague's involvement in analysis, decision-making and activity implementation is critical to consolidate his reliability. The driver must be available to back up the colleague in higher risk activities to avoid to put the stake in danger and to reduce the stress of the colleague.
MBO features	Type of goals to be set	Tasks	To optimize the performance of the colleague, the driver subdivides his action into *tasks (sets of steps),* whose level of complexity is higher enough to challenge the colleague while being consistent with his competencies (Blake and Mouton, 1985).
	Type of control to be carried out	Focused	To recognize colleague's reliability and foster his commitment, the driver controls only the activity's key factors of success. In case of mistake he provides reorienting feedback and reacts rapidly to help the colleague fix the issue.

Table 3.3.12—Features of the relationship between the driver and the colleague with regard to a sufficient level of autonomy

Level of autonomy of the relationship		*High*	A low stake activity has to be carried out by a colleague who has a high level of reliability, which is originated by one of his logics of functioning: interdependence, pseudo-dependence facing a simple activity, independence facing a challenging activity.
Drivership style (aimed to develop colleague's reliability)		Empowering	*Sharing power and responsibilities* with the colleague and providing him with appropriate means (human, financial, material…) to implement the action and protect the stake
Role the colleague expects the driver to play		Mentor	*Orienting* the colleague's activity, *asking questions* to stimulate his reflective thinking, *providing feedback* to improve his understanding of the context and his stability in decision making and action implementation
Main colleague's guidance need		Orientation	Because of the low stake and colleague's high level of reliability, the driver has just to orient his action by making explicit the global desired outcome and being available to discuss the major issues the colleague faces in his activity.
MBO features	Type of goals to be set	Mission	The driver gives the colleague a *mission (set of tasks)* whose level of complexity is high enough to challenge the colleague's commitment and competencies (Blake and Mouton, 1985).
	Type of control to be carried out	Random	When the mission is completed, the driver checks the outcome by randomly controlling some of the colleague's actions to make sure that things have been done as agreed and in compliance with internal and external regulations.

Table 3.3.13—Features of the relationship between the driver and the colleague with regard to a high level of autonomy

Implementing the Autonomy Curve

Autonomy must be customized to the context of the relationship. Different activities carried out in the same context (or the same activity carried out in different contexts) can require different levels of autonomy in the relationship between the driver and the colleague. In fact, autonomy depends on as many contextual factors as the impact the activity has on the driver's stake, the colleague's level of commitment and competencies, the relationship the colleague has with the others actors in the context (Blake and Mouton, 1985). In the contextual drivership paradigm, different levels of autonomy may be required in the relationship between the driver, each member, and the team as a whole. It may happen that the relationship between the driver and the team requires a high level of autonomy while the relationship between the driver and some of the members of the team requires a low level of autonomy. The contrary can be true.

Assessing the appropriate level of autonomy in the relationship is a process aimed to facilitate and to foster the sharing of own perceptions between the driver and the colleague (Porter and Lawler, 1968; Lax and Sebenius, 1986, Lee and Koh, 2001). When discussing autonomy, people should make reference to facts (Hersey, 1984).

Let's make an example of an activity where the relationships starts at a low level of autonomy and evolves up to a high level of autonomy.

At the beginning, the driver allocates most of his time to properly organize the Colleague's activity. The trade-off is made by reducing the time allocated to communication, which is "quantitative," factual, and explanation-oriented (Blake and Mouton, 1985). Since success increases commitment and learning increases competencies (Watzlawick, Beavin, and Jackson, 1967), to start enhancing the colleague's level of autonomy, the driver must:

- Create the material context that allows the colleague to perform the activity in the best possible way;
- Reduce his anxiety with regard to failure by structuring the activity;
- Foster the learning process by making quick success happen (Kotter, 1995, 1996).

By that time the colleague's reliability increases, allowing the relationship to shift toward a limited level of autonomy. The colleague is able to perform the activity in a still basic but more secure way and starts asking questions and looking for a better understanding of the context of the activity. The increase of his reliability allows the driver to decrease the time allocated to the organization of the colleague's activity and to increase the time allocated to the communication with him. Such a change has a positive impact on a colleague's commitment, as well as his competencies. At this stage it is important that the driver allocate more time to follow up the colleague in his activity, support the activated virtuous circle, and further support the increase in the colleague's reliability.

By this time it is possible to shift toward a sufficient level of autonomy in the relationship. In fact, since the colleague's reliability has increased, he is able to organize most of the activity by himself, and he is ready to propose his own solutions. The colleague needs more time to exchange his ideas with the driver and less help in the execution of the activity. The driver starts focusing more on "qualitative" communication by reducing the time allocated to the communication with the colleague while being available the colleague should need help. By behaving in such a way, the driver provides a positive reinforcement to the colleague (Hersey and Blanchard, 1972), who feels encouraged to go in the process of increasing his reliability.

When the level of autonomy becomes high, the colleague no longer needs help in the execution of the original activity, but needs guidance to optimize his performance. The communication process is then very much focused on the overall purpose of the activity for the colleague to protect the stake of the driver by properly carrying out the activity by him. The driver can then allocate most of his time to other colleagues or other activities of the same colleague, in so doing, reinforcing a virtuous circle of commitment and competencies improvement.

Follow-up Protocol (FiP) and Support Protocol (SAS)

The follow-up protocol (FIP—Fixed Points) and the Support Protocol (SAS—Slot of Active Help) help the driver to fully leverage the autonomy curve by adapting his drivership style to the colleague and by fine-tuning their relationship to the evolution of the context. If the FiPs are satisfactory and the colleague requests few SASs, the driver may induce that his drivership style is adapted to the colleague's reliability and to the stake to be protected; the driver may even consider to increase the level of autonomy of the relationship with the colleague and therefore modify his drivership style by shifting right on the autonomy curve. If the FiPs are satisfactory but the colleague requests too many SASs, the driver may induce that his drivership level is adapted to the colleague's reliability, but the colleague still needs support when facing difficulties. In this case, the driver should not modify his drivership style, and should keep a closer contact with the colleague to help him acquiring self-confidence.

If the FiPs are not satisfactory (and if the SASs have been duly carried out by the driver when requested by the colleague), the driver may induce that his drivership style is not appropriate and therefore modify it by shifting left on the autonomy curve. However, the key factor of success is the way the driver carries out the follow-up of the activities; that is, how he behaves when controlling and supporting the colleague. The following paragraph provides the FiP and SAS protocols useful to the driver to carry out these two activities in the most effective and efficient way.

The FiP protocol is divided into four main phases, as shown by Diagram 3.3.4 and Table 3.3.14. Table 3.3.15 shows what the interests are for the driver and the colleague in properly carrying out a FiP.

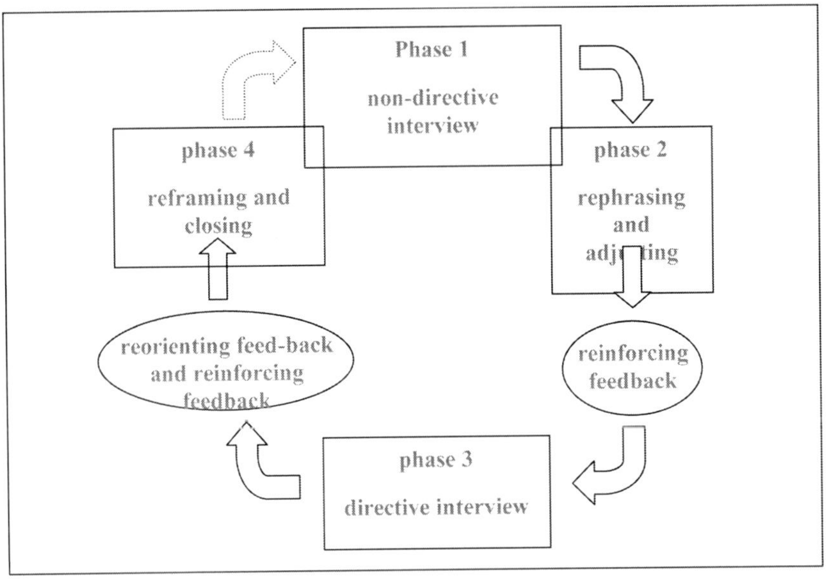

Diagram 3.3.4—Process to implement a fixed point

To orient the colleague's action while checking the colleague's understanding of the origins of the issues, in questioning, the driver should pay attention to find the appropriate balance between the "how did you do" approach and the "why did it happen" approach (Watzlawick, Beavin, and Jackson, 1967).

phase 1 **non-directive interview**	In this phase, the driver - Asks the colleague to tell *how* he carried out the assigned activity and lets the colleague free of answering in the way he prefers; - Keeps silent and, if necessary, asks open questions; - Listens actively and pays attention to the whole patterns of communication (verbal, para-verbal, non-verbal) used by the colleague, in order to collect as much information as possible.
phase 2 **rephrasing and adjusting**	In this phase, the driver - Rephrases what said by the colleague to confirm him that the messages have been received correctly; - If the colleague considers he needs to rectify or restate some of his sentences to clarify what he meant, he is free to do so; At the end of phase two, the driver gives a reinforcing feedback to the colleague. This reinforcing feedback is necessary to legitimate further reorienting feedback
phase 3 **directive interview**	In this phase, the driver - Focuses on the issues that he thinks are critical and on which he needs more details; - Conducts the interview by asking closed questions; - Matches information gathered in phases one and two with the answers given by the colleague in phase three. At the end of phase three, the driver gives a two steps feed-back to the colleague, starting with a reorienting feedback on the activities that he considers have not been carried out in a proper way by the colleague; then he gives a reinforcing feedback on the activities that have been carried out in an appropriate way by the colleague.
phase 4 **reframing and closing**	In this phase, the driver - Provides a global assessment of the colleague's performance; - If necessary, restates the rules of functioning and reframes the activity of the colleague; - Sanctions the colleague with rewards or penalties; - Adjust own drivership style according to the performance of the colleague; - Briefs the colleague on the next activity to be carried out.

Table 3.3.14—Features of a fixed point

Phase	Driver's interests	Colleague's interests
1	- To collect as much information as possible; - To have a general overview of the appropriateness of the level of autonomy by observing the way the colleague debriefs his activities.	- Freedom to deliver the messages in the way he prefers; - Possibility of acting according to own level of autonomy;
2	- To demonstrate to the colleague that he is actively listened and his action is fully considered; - To make sure that all the information provided by the colleague have been clearly understood; - To reward through the reinforcing feedback colleague's efforts and positive performance	- Possibility of rectifying the information delivered to avoid misunderstandings; - To be recognized in his efforts and performance through the reinforcing feedback
3	- To check the consistency of information gathered in phases one and two, and to go deeper into specific points that have not been duly covered by the colleague by to focusing on factual information; - To make necessary corrections to avoid drifts in the colleague's activities through the reorienting feedback, and to reinforce colleague's virtuous behaviors through the reinforcing feedback	- To provide further details and facts on specific items; - To get specific instructions on how increase own performance through the reorienting feedback - To understand driver's main interests in order to anticipate them in the next phases
4	- To give feedback to the colleague on his performance and behaviors; - To reinforce the implementation of the ground rules through rewards and penalties; - To launch the next phase of the job to be carried out and to redefine the type of drivership style to be adopted	- To understand driver's perception of colleague's performance; - To be recognized for virtuous behaviors and performance through rewards or to be reoriented through penalties; - To have the possibility of discussing with the driver the level of autonomy in their relationship adapted to colleague's reliability

Table 3.3.15—Driver and colleague's interests in implementing FiPs

Support Protocol: the Slots of Active Support (SASs)

The SAS protocol is designed to help the driver and the colleague optimize the time and resources allocated. A SAS has to be carried out by making reference to the level of importance and urgency of the activity as shown in the time management matrix (see Table 3.3.4).

A SAS is conducted adopting a "how to solve the issue" approach instead of a "why did it happen" approach (Watzlawick, Beavin, and Jackson, 1967), as shown by the basic guidelines explained in Table 3.3.16.

Main points	Rationale
Prior to any action of support driver checks whether the colleague has or not identified one or several possible solutions.	*It helps the driver to check if the level of Autonomy of the relationship with the colleague is correct.* If the colleague's proposal is acceptable, then the driver has just to support him in decision-making.
If the colleague has no solution, the driver checks what is the level of urgency of the issue before going on with the SAS	*It helps the driver to optimize his time.* If the issue is not urgent, than it is possible to ask the colleague to look for a solution. If the issue is an urgent one, then the driver must allocate time (or find somebody) to help the colleague identifying a suitable solution.
If the colleague has a solution, the driver checks what is the level of importance of the issue before going on with the SAS	*It helps the driver to optimize the performance of the solution.* If the issue is not important, then he can validate the colleague's solution without further inquiry. If the issue is important, then the driver must allocate time (or find somebody) to analyze and adapt colleague's solution.

Table 3.3.16—Basic guidelines of a SAS

The decision tree presented in Diagram 3.3.5 shows the four possible patterns in conducting a SAS.

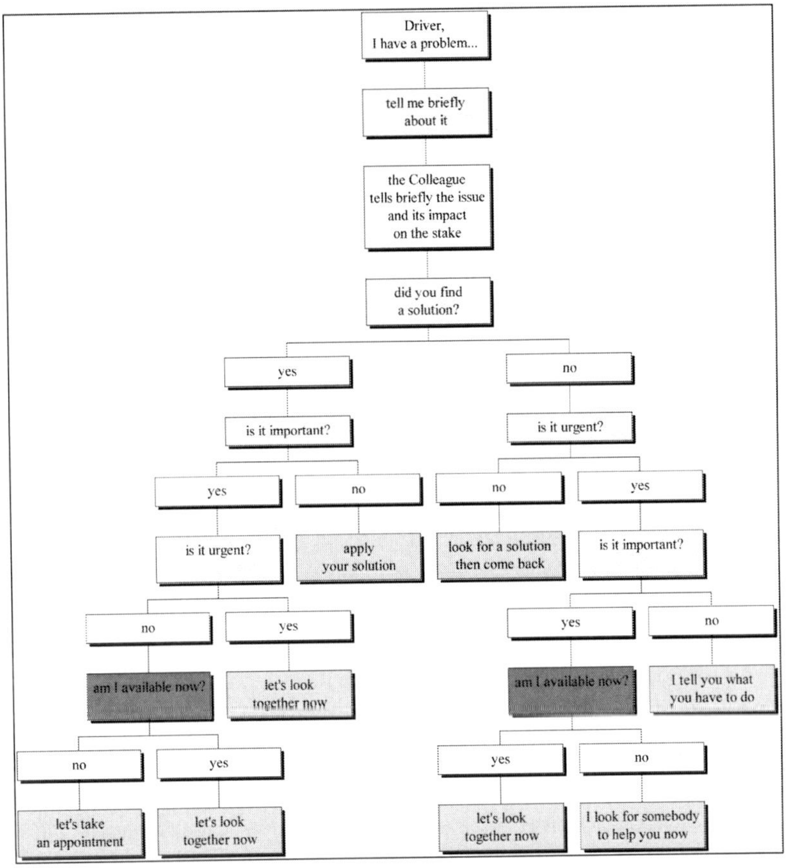

Diagram 3.3.5—SΛS protocol

3.3.5 Strategies to Build Power Patterns

The drivership model assumes that a "context" is a contrived human system which can be understood, based on the interaction of its elements/parts. A context is sustained through repetitive behaviors and power relationships, which often are habituated and, therefore, below the individual's and/or group's level of awareness.

For change to happen, a driver must be able to deconstruct the context, identify the behaviors that sustain the existing context, present other appropriate behaviors that are seen as stake-supporting, and develop a power relationship based on expertise and perceived commitment to the well-being of the individual(s) and organization.

The Adaptability Matrix

The adaptability matrix (Table 3.3.17) integrates the driver's capability in understanding the context and modifying his behaviors.

		Behavioral flexibility	
		Low	High
context's understanding	High	*Resistant* The driver knows what behaviors are required by the context, but is unable to adopt them. He is conscious of his mistakes, but does not know how to correct them	*Adaptable* The driver is able to adopt the behaviors that are the most appropriate to the context. Trained to self-mastery, he feels comfortable in any context
	Low	*Rigid* The driver adopts the same behavior regardless of the context. He is not conscious of his behaviors nor of the evolution of the context, he does not accept to question himself	*Unpredictable* The driver adopts various behaviors who have no relationship with the context. He generates stress and interpersonal conflicts within the change process

Table 3.3.17—Adaptability matrix

To correctly implement the autonomy curve protocol, the driver must adopt behaviors that are appropriate to the features of each colleague and of the team as a whole. Adaptability, a form of tacit knowledge, is embedded in action and necessary for predictable proficient managerial performance (Polany, 1962; Nelson and Winter, 1982)

Within a context, actions are taken that may be open to interpretation unless the underlying meaning is articulated. Once articulation occurs, the process activities, which can sustain knowledge creation, can be identified and instituted (Spender, 2004). "New" knowledge that supports the change initiative and outcomes may be an extrapolation of existing knowledge or idiosyncratic knowledge, which replaces tacit knowledge. If the former, slight alterations in habituated behaviors may be sufficient. If knowledge is idiosyncratic, cognitive re-framing is necessary for the acquisition and habituation of the "new" behaviors.

Stretching Penrose's perspective (1959), an accurate understanding by the driver of a context incorporates an awareness of the systemic links and relationships that exist among the elements of a system. The understanding may allow the driver to leverage opportunities, which emerge as derivatives of the change process and create value for the firm (Spender, 1996, 2004).

The Power Patterns Matrix

Hierarchy is in charge of ensuring adherence to the agreed ground rules and regulations in order to maintain stability. Change can deconstruct existing power patterns and bring into question the efficiency of existing rules and regulations.

The construct drivership is developed to overcome the intertwining of the hierarchical assigned role responsibilities with the constructs of leadership and management. Induced change is resisted to the extent it is not linked to an acceptable stake, the individual's interests in achieving the stake is not articulated, the existing position power relationships are altered without acceptance of such modifications, and behavioral change is mandated by hierarchy instead of being developed by learning.

Behavioral flexibility is more likely to develop if people are able to experiment with new behaviors in their daily practices. In a context where people are free to explore new practices, allowed to make mistakes, encouraged in their reflective thinking to correct them, and supported in sharing their discovery, behaviors supportive of the change process and desired stake are more likely to become routine practice. Therefore, the driver must engage in acquiring and leveraging power, which is accepted as legitimate within the change context. From this point of view, the enactment of an effective, cooperative change process requires the driver to appropriately leverage power to make possible the emergence of effective practices useful to protect the stake and achieve organizational goals (Spender, 2004).

By making reference to the most common power taxonomies (French and Raven, 1959; Etzioni, 1961) within the paradigm of drivership it is possible to identify two types of power:

- Position power makes reference to legitimate coercive, connection, and/or reward power of the individual.
- Personal power makes reference to the legitimate expert, information, and/or referent power of the individual.

Regardless of whether the driver has position or personal power, he needs to have the interpersonal skills to be able to have the colleague perform the activities necessary to protect the stake (Spender, 2004).

Traditional position power is earned through a process of adhering to the norms established by the existing power structure (promotions happen often because the promoted individual is perceived as supporting the rules, regulations, processes, and procedures established by the hierarchy). When change is introduced, an individual is designated as "the" change agent and sometimes granted power of position by the "superiors." Those above and those below and/or on the "same" level may have different expectations as to what are desired change "behaviors." Change power may not be granted by those who must "change."

If the change agent is astute, he understands the need to develop personal power to alter any negatives assigned to the change agent position by the changed. In addition, the change agent may have to help those above alter their behaviors and cognitive frame if the change initiative is to succeed.

For the drivership model, an integration of position and personal power is essential. A power pattern is then the result of the interdependent use of these two types of power, as shown in the Table 3.3.18.

		Use of personal power to leverage influence and relationship	
		Low	High
Use of position power to put pressure on output achievement	High	*Authoritarian Power Pattern* The focus on colleague's compliance with driver's decisions and on the fulfillment of the defined objectives is higher than the focus on driver's relationship with the colleague	*Systemic Power Pattern* The focus on colleague's compliance with driver's decisions and on the fulfillment of the defined objectives is equal to the focus on driver's relationship with the colleague
	Low	*Stepping Back Power Pattern* Feigned indifference to both colleague's compliance with driver's decisions and established objectives, and driver's relationship with the colleague	*Contractual Power Pattern* The focus on driver's relationship with colleague's is higher than focus on colleague's compliance with driver's decisions and on the fulfillment of the defined objectives

Table 3.3.18—Power Patterns Matrix

The contextual analysis and community norming phases address the issues related to the hierarchy role and the allocation of the position power between the actors of the change process. These processes reference the legitimacy of each actor to identify the community's stake, the enactment of ground rule construction and interface-setting, the enforcement of ground rules, and, when necessary, the arbitration of conflicts in ground rule interpretation.

The contextual drivership and the cooperative innovation phases address the issues related to the leadership/management roles, the earning of personal power through interpersonal interactions among the actors of the change process. These constructs reference the legitimacy of each actor to shape the relationships with the others to "control" the purpose of the change, make decisions about the implementation and the outcome of actions, which protect the stake, and leverage the available competencies for identifying appropriate solutions and enhancing best practices.

Mastery of the four power patterns allows the driver to focus on the uniqueness of specific relationships. This enhances driver's influence and is useful to diminish colleagues' resistance in the change process. In fact, an adaptable driver is able to strategically leverage several power patterns to propose acceptable meaning, evolving the organizational routines, creating control, and stimulating synergies by enhancing the sharing of mutually consistent practices (Spender, 2004).

In a cooperative change process, the systemic power pattern is preferred because it nourishes relationships, which support compliance and goal achievement. At the same time, goal achievement supports relationship. The outcome of such a power pattern is the implementation of successful and effective drivership practices (Bass, 1960), and it should be used as often as possible.

However, not all change processes are cooperative, and even cooperative change may experience difficulties. In these contexts and instances, the adaptable driver recognizes that one or more power patterns may be necessary.

The stepping back power pattern is particularly useful to solve stalemates, although the choice of using no or little power is sometimes considered a non-leadership behavior (especially in western philosophy and management literature). The stepping back power pattern is based on the perspective that "facts are more stubborn than people" (Antonio Gramsci). The driver lets events happen that he believes are helpful in having the counterpart begin to listen and accept another's point of view.

Stepping back does not mean the driver gives up control. Instead, the driver allows the colleague to learn appropriate ways of acting because of the confrontation of reality. The stepping back power pattern is particularly useful with individuals so convinced of the correctness of their approach that they cannot consider any another viewpoint (Hersey and Blanchard, 1972). If their actions do not endanger the stake or the driver's main interests, then discovering the inappropriateness of their behavior has a lasting effect. However, if an individual's actions are detrimental, another power pattern should be used.

For aggressive people who tend to put themselves in a one-up position to impose their own point of view, stepping back can destabilize because the aggressor expects resistance. In addition, others may be persuaded to not accept such behavior, because it disrupts achieving the stake. If others committed to the stake intercede, the driver can stay in a one-down position, which avoids a symmetric escalation of the confrontation (Watzlawick, Beavin, Jackson, 1967; Malarewicz, 2000). Stepping back is a difficult skill to master because it requires a driver functioning at the highest levels of adaptability, context analysis, self-awareness, and interpersonal skills. The stepping back power pattern should not be used when the difference in power becomes too high and could lead to domination, or when it could give rise to laziness.

Contractual power is useful to establish and consolidate relationships with higher hierarchical individuals or those relative independent from hierarchy who cannot be influenced by position power, e.g., members of an interdisciplinary cross-functional project team. In this context, the project manager may use contractual power to link functional competence to specific activities and thus acknowledge an individual's value. Then the manager can gradually shift toward systemic power, having established the legitimacy of his position. A risk of contractual power is failure to enforce compliance and fulfillment of the defined goals. Systematic, periodic feedback on performance and progress must be built into the contract in conjunction with consequences.

The authoritarian power pattern should be used sparingly because it can provoke interpersonal conflicts and harm the relationship in the long term. However, in two situations, the driver should not hesitate to shift to this power pattern:

- When the rules of conduct are repeatedly not respected notwithstanding several warnings. In this case, the tit-for-tat strategy, described in the previous chapter (Axelrod, 1984), can be used briefly to re-establish the legitimacy of the driver, the stake and the change activities;

- When the context requires a decision concerning a matter of safety or compliance with laws belonging to the core activities of the driver and affecting directly his stake.

The authoritarian power pattern should always be followed by the adoption of the systemic power pattern in order to reconstruct the relationship with the counterpart.

A driver must use power differentially based on the context, the functioning of each individual, and the team as a whole. It may happen that the driver chooses to follow a systemic power pattern (or any other of the four) with the team while adopting an authoritarian power pattern with some members of the team, a contractual power pattern with some others, a stepping back with others, and a systemic power pattern with others. The aim of such a strategy is for the driver to adapt as much as possible his behavior to the context of the team and to the features of each individual. However, the final target of any power strategy should be the systemic power pattern.

3.3.6 Commitment Enhancement Matrix

Under the contextual drivership paradigm, commitment plays a central role to assess individuals' and team's reliability and to define the level of autonomy in the relationship. The main task of a driver is to create an atmosphere within the organization that elicits a committed attitude and initiates harnessing of a colleague's energies in the right direction (Ouchi and Jaeger, 1978; Ouchi and Johnson, 1978). This means that a driver cannot motivate his colleagues but can create a favorable working context that may activate inner intrinsic factors leading the worker to feel motivated (Watzlawick, 1983; Tissier, 1988). Employees' attitude determines success at work and can make a huge difference among employees. Matching job features and workers' expectations enhances workers' satisfaction (Herzberg, 1959). Hygiene and motivation factors must be activated simultaneously to enhance workers' performance. This means

- Treating people as best one can so they have a minimum of dissatisfaction;
- Making sure that people can get achievement, recognition for achievement, interest, and responsibility to grow and advance in their work.

Commitment and Motivation in the Contextual Drivership Paradigm

The commitment enhancement matrix combines the comfort dimension and the stimulation dimension, the two factors that most impact people's commitment. The matrix provides insights on how a driver can create an appropriate working context to foster people's commitment and to avoid its dropping.

Motivation is the result of intrinsic drivers activated by appropriate extrinsic contextual factors, which leads an individual to satisfy his own needs/interests and to pursue organizational goals by doing things when he wants to do them and by doing them without constant prompting by other people (Vroom, 1964; Tissier, 1988; Deci, 1992).

Commitment is a state of being of an individual that is grounded on his awareness of the context. No matter what satisfiers and dissatisfiers factors are present in the context, commitment leads the individual to explicitly and voluntarily leverage his own physical, intellectual, emotional, and intuitive resources to protect his own and his team's stakes and contribute to his own and his team's actions and accomplishments (Herzberg, 1958, 1966; Becker, 1960; Buchanan, 1974; Porter, Steers, Mowday and Boulian, 1974; Salancik, 1977a, 1977b; Wiener, 1982; Meyer and Allen, 1997; Kanter, 1968).

The concepts of commitment and motivation are apparently quite similar, but from the drivership perspective, the former is more valuable.

Commitment represents employees' sense of identity with the enterprise and its mission (Nonaka, 1991; Hebbar, 2003). A committed colleague decides to leverage his resources because he is aware of his contribution to the community and consciously wants to respect the moral contract passed with other people who are linked to him (Senge et al., 1994; Allen and Meyer, 1990). Comparing committed and motivated colleagues, the former resists to the worsening of the context for a longer than average period than the latter, because he tries harder to eliminate the obstacles instead of giving up when they face difficulties. When hygiene and motivation factors decline, the former continues to leverage his best resources and goes along in producing the best efforts to provide effective contribution to team's actions. The latter, when extrinsic factors weaken or disappear, reorients his resources toward other tasks that better satisfy his new own needs (Zeigarnik, 1927; Blake et Mouton, 1985).

Using commitment instead of motivation can be useful for several reasons. Commitment requires the colleague to explicitly say what are the actions and accomplishments he binds himself to. It helps the colleague to clarify his relationships and roles in the team and look for a common stake to be protected, instead of driving individuals toward the satisfaction of their own needs or interests (Salancik, 1977a, 1977b; Allen and Meyer, 1990).

Commitment-driven change processes show a rate of success higher than those driven by conviction (Joule et Beauvois, 1987, 1998), because individuals adapt their beliefs to make them consistent with their acts (Cialdini, 2001; Piattelli Palmarini, 1995), and change is the outcome of the implementation of colleagues' ideas rationalized, verbalized, and put into actions by themselves (Watzlawick, Weakland, and Fish, 1974).

To leverage commitment, the driver has to make sure that the colleague:

- Verbalizes his feeling and his own understanding of the context;
- Recognizes his own, others,' and common stakes;
- Describes what are the satisfiers and dissatisfiers factors enacting his actions;
- Explicitly identifies the competencies he can use or must develop to comply with his promise;
- Acts accordingly to what was verbalized in the previous steps to link actions and accomplishments.

The commitment enhancement matrix (Table 3.3.19) combines the two dimensions that a driver should consider to design an appropriate working context: the comfort (dissatisfier factors) and the stimulation (satisfier factors). Each dimension integrates the contributions of Lewin (1950, 1951), Maslow (1943, 1968), Herzberg (1959, 1966), Vroom (1964), Porter and Lawler (1968), and Salancik (1977a, 1977b).

The comfort dimension makes reference to the factors impacting the colleague's working context from a material point of view: Herzberg's dissatisfiers (hygiene factors), Maslow's physiological, safety/security, and belongingness needs. Setting the comfort factors at the appropriate level can reduce and sometimes eliminate job dissatisfaction because it fulfills basic human needs. The absence of such factors in the working environment can trigger commitment decline while their activation has just little positive effect on commitment improvement (Johnson and Jones-Johnson, 1992).

		Not appropriate	Appropriate
Comfort	Appropriate	*Short-term Commitment Stagnation* In a high-comfort/low-stimulation context, commitment stagnates. People do not experience job dissatisfaction because their expectations concerning basic material needs are satisfied. People avoid expressing negative feelings because they do not have anything to complain with of. However, because of the absence of stimulation factors, people's more advanced needs are not fulfilled and their commitment to work cannot increase. Psychological growth is inhibited (Herzberg, 1966), and the context is stable only in the short-term. To keep such stability, the level of comfort must steady increase, because people continuously raise their expectations concerning the comfort at work. If such an increase in expectations is not met, the context gradually skids toward the southwest cell; that is, conflict or resignation.	*Long-term Commitment Reinforcement* In a high-comfort/high-stimulation context, people are more likely to increase their level of commitment because their expectations are met. Psychological growth is allowed and encouraged (Herzberg, 1966) because the working context is secure, challenging, and rewarding. Such a context is often stable and contributes to commitment enhancement in the long-term, because of the virtuous circle created by the mutual reinforcement of both dimensions. The level of stress is low, people adopt proactive, creative, positive attitudes to work, and are able to meet more ambitious goals.
		Long-term Conflict or Resignation In a low-comfort/low-stimulation context, people are dissatisfied with their working environment and do not find satisfaction in their job, and usually adopt one of the following two attitudes to work: - Conflict: people fight against such a context and look for better working conditions through an open conflict (increase in salary, improvement in environmental ergonomy, decrease in stressing interpersonal relationships, actions against moral harassment etc.) - Resignation: people invest their energies in activities not directly related to the job, adopt passive behaviors, avoid creative contribution, do not take risks. In both cases the observed outcomes are (Porter and Lawler, 1968): - Increase in personnel turnover, unexplained absenteeism, work accidents and injuries, quality defects, waste, clients' complaints, suppliers' disputes, conflicts (interpersonal, intra-organizational, inter-organizational), hidden costs, etc.; - Decrease in productivity, innovative suggestions, time keeping, resistance to workload, etc. In such a context, commitment to work tends to decrease rapidly, and some radical changes are necessary to ensure an effective turnaround. Sometimes, the only possible solution is a radical change of the management teams, with all the consequences such a decision can have on the life of the company.	*Short-term Endurance to Exceptional Events* A low-comfort/high-stimulation context is typical of exceptional situations where mobilization is required to face a danger or a special event, most often coming from outside. Such a context can be originated by - An unforeseen degradation of the comfort dimension, when the context is initially in the north-west cell. If people are properly informed of what happens or is going to happen and are convinced that such comfort worsening is temporary, no reduction is observed in the level of commitment. On the contrary, sometimes it can even increase, because of the will of people to make all efforts to fight against external factors. - A sudden improvement of the Stimulation dimension, when the context is initially in the south-west cell. In this case, people accept to leverage their resources because some important emotional drivers are enacted, either directly by the external event or by the driver who is able to use the external event for this specific purpose. The conflict or resignation attitude existing before the event will even worsen after the end of the mobilization if nothing is done to improve the comfort dimension, or if the stimulation dimension is forgotten again. In both cases, people can demonstrate endurance to exceptional events only for a reasonably short period. How short is short depends on the context, on the event and on the individuals. If the comfort dimension does not improve, the outcome is always conflict or resignation.

Table 3.3.19—Commitment enhancement matrix

A driver can utilize the commitment enhancement matrix to identify patterns of evolution of a colleague's attitude at work and gather ideas to create a positive working context.

Production output is influenced more by the type of relationship that the worker has with his supervisor than by the type of environment he works in. Therefore, the contextual driver customizes as much as possible his behaviors and the material working conditions to the features and expectations of each colleague.

The proper use of the autonomy curve can help in adapting the relationship to foster colleague's commitment to work. The systemic nature of contextual drivership provokes a virtuous circle leading to performance increase, autonomy increase, and commitment increase (O'Reilly, Chatman and Caldwell, 1991).

The driver has to make sure that the comfort dimension is higher than his colleagues' expectations. The worsening of working conditions can provoke:

- In the short term, an increase in the level of stress that people experience in their day-to-day working activities;
- In the long term, a negative impact on the effectiveness of the factors enacted in the stimulation dimension, and thus a shift of colleagues' attitude toward conflict or resignation.

Many factors contribute to the comfort dimension; the contextual driver must be able to identify, choose, and combine them to get the comfort always above the minimum expected level by using the lower amount of resources. However, people's expectations concerning the comfort dimension have a downstream inertia higher than upstream inertia: people claim regularly for continuous improvement in the working environment, and once a given level of comfort is achieved, this very level is considered as acquired forever.

On the other hand, colleagues are usually not keen to give up an acquired level unless the context is dramatically modified by unforeseeable events. Therefore, it is important to highlight the fact that salary impacts the comfort dimension more than the stimulation dimension.

Any increase in salary (even production bonuses when they are contractually set in advance) is quickly discounted by the colleague as a standard feature of the comfort dimension. Only money associated to recognition for an achievement really impacts the stimulation dimension (Herzberg, 1959, 1966).

When employees can play a role in the setting of goals and the making of decisions affecting their work, they accept change more readily than those who are catapulted into change without any opportunity for the exercise of free choice (Coch and French, 1948).

The cohesiveness of a group is a key factor to control the behaviors of its members. (Lewin, 1950, 1951). To set the commitment dimensions at the appropriate level, a contextual driver continuously collects information by observing, questioning, and being factual on events.

He feeds and supports the stimulation dimension through frequent positive feedback, appropriate career planning, and intelligent job enrichment. Examples and achievements are powerful drivers to enhance colleagues' commitment to work. Fair procedures, behaviors, and sanctions (rewards and penalties) help people to focus their energies and resources to improve their performance.

When something unexpected happens, the contextual driver provides people with reliable information, is open to suggestions, and is clear and firm on the stake to be protected and the goals to be met. Nothing can badly affect commitment as lack of transparency in driver's actions.

3.3.7 Competencies Development Cycle Matrix

The competencies development cycle matrix (Table 3.3.20) is the result of the integration of Herzberg's (1966) job enlargement and its derivative job rotation and job enrichment, Handy's (1989) portfolio activities approach, Aubrey's (1994) discourse on employability, and Nonaka and Takeuchi's (1995) studies on knowledge creation steps. The competencies development cycle matrix is the combination of two dimensions:

- The weight of each assignment compared to the total time allocated by the colleague to his professional assignments (the frequency dimension);
- The contribution (perceived by the colleague) of the assignment to his professional and personal development and the improvement of his competencies (importance dimension).

		Frequency of the assignment in the individual assignments portfolio	
		Low	High
Importance of the assignment for the professional and personal development of the colleague	High	*Development Assignments* Assignments based on the tacit knowledge stored in the sleeping-base and carried out to acquire new distinctive competencies and to satisfy future core interests	*Key Assignments* Assignments belonging to the core mission, carried out to satisfy critical interests by leveraging competencies acquired in development assignments phase
	Low	*Sleeping-base Assignments* Assignments carried out in the past and no longer used, or marginal activities satisfying no core interests	*Routine Assignments* Standardized assignments carried out by using competencies that can no longer be improved or leveraged to satisfy major interests

Table 3.3.20—Competencies development cycle matrix

Changing periodically the mix of their assignments portfolio (Handy, 1989) helps individuals to protect their employability (Aubrey, 1995), because in this way competencies are kept up to date by the appropriate use of the existing ones, the development of new ones, and the dropping of those that are no longer useful. This can be compared to the product life cycle:

- At an early stage competencies must be discovered, explored, and acquired through a controlled investment (the development assignments phase corresponds to the product development phase);
- As soon as these competencies become distinctive and can provide increasing added value, they must be intensively leveraged (the key assignments phase corresponds to the product growth phase);
- When these competencies are sufficiently mastered and the potential for improvement and increase in performance has decreased to a stable low level, they must be used to minimize the use of resources to produce stable added value (the routine assignments phase corresponds to the product maturity phase);
- As a final stage, obsolete or low added value competencies must be abandoned to make resources available for the development of new edge competencies (the sleeping base assignments phase corresponds to product decline phase).

The Process to Facilitate Assignments' Rotation

The appropriate use of the competencies development cycle matrix (CDCM) helps the driver set up a learning organization by encouraging individuals and teams to experiment new approaches, learn from others' experience and best practices, and transfer knowledge quickly and efficiently throughout the organization (Garvin, 1993). The CDCM ensures and facilitates the clockwise rotation of the assignments in the four cells of the competencies development cycle matrix, as shown in Diagram 3.3.6.

The first step is to identify all colleagues assignments, prioritize them by taking into account the drivers,' company's, and colleagues' interests and needs, and then plot them into the matrix. To start the rotation, it is necessary to make a first move; that is, to free time in one of the cells and allow a redeployment of the assignments in the others. Nonaka and Takeuchi's knowledge-creation cycle (1995) is helpful to identify the way to make it rotate, as shown in Diagram 3.3.6. The following paragraphs show in detail how to make the process work.

Transferring Routine Assignments in Sleeping base Assignments

To increase the size of the subsidiary activities or the scenarios activities cells in the driver's added value matrix, each colleague has to identify his core activities and transfer to others the assignments with lower added value. Other colleagues have then to acquire the competencies necessary to execute such transferred assignments.

To activate the assignments' rotation process in the competencies development cycle matrix in order to move up the "do or make do line" in the driver's added value matrix, the colleague has to:

- Identify those routine assignments that he carries out in a standardized way;
- Make explicit the competencies necessary to carry out these routine assignments;

- Transform these assignments (and the related competencies) in sleeping base assignments by transferring them to another person, for whom they will become development assignments.

The repetitive use of explicit knowledge has extended the colleague's tacit knowledge and activated the internalization process of the routine assignments transformed into sleeping base assignments. To leverage them again, he will just need to spend some time to recall them and to recover the level of mastery he had when they were in his key assignments or routine assignments (Nonaka and Takeuchi, 1995).

Consolidating Key Assignments into Routine Assignments

To increase the size of the subsidiary activities cell and the core activities cell in the managerial added value matrix, the "certain or uncertain line" must shift left. This implies increasing the possibility of planning the activities by standardizing them.

Transferring some routine assignments to other people allows the colleague to consolidate some of his mature key assignments into new routine assignments, which will be carried out using the resources made available by the previous transfer.

Rotation helps the colleague look at the business from multiple perspectives and understand the importance of transferring assignments to others (Nonaka, 1991). It allows the driver to modify his drivership style by increasing the level of autonomy in his relationship with the colleague. This corresponds to the combination process; that is, the standardization of explicit knowledge into public available protocols (Nonaka and Takeuchi, 1995). The purpose of such process is to make sure that the colleague combines his own explicit knowledge with other available explicit knowledge to standardize it and make it ready to be transferred to others.

It allows the company to build redundancy and facilitate the sharing of tacit knowledge through the strategic rotation of people into different positions and the creation of a common cognitive ground among employees. The cumulative effect of the previous two steps is to increase the global influence of the subsidiary activities cell and the horizontal flattening of the core activities cell. The consequence is an increase in the efficiency of the driver and his effectiveness.

Formalizing Development Assignments to Transform Them into Key Assignments

The resources made available by the standardization of key assignments into routine assignments can be invested in formalizing and leveraging the competencies acquired in the development assignments (low frequency/high importance) to make them becoming new key assignments. This corresponds to the articulation process; that is, the translation of tacit knowledge into explicit knowledge (Nonaka and Takeuchi, 1995). Formalizing the competencies helps to identify which are the most suitable development assignments to be transformed into key assignments.

Transforming development assignments into key assignments must correspond to a shift from limited to sufficient in the level of autonomy of the relationship between the driver and the colleague.

The articulation of the tacit knowledge and its formalization must be properly monitored by the driver according to what is proposed by the autonomy curve. Implementing key assignments without an appropriate control of their outcome and regulation of the intensity of the workload can increase the risk of burnout for the colleague. Intensive early development (reasonably high workload concentrated in short time periods) and constant increase in professional challenges (Quinn, Anderson, and Filkenstein, 1996), coupled with the appropriate use of reflective thinking (Senge et al., 1994), are necessary to enhance the efficiency and the effectiveness of distinctive competencies characterized by a high degree of growth potential.

Identifying Development Assignments in the Sleeping Base

When a development assignment is transformed into a key assignment, the colleague must identify other development assignments to acquire new competencies and prepare the next rotation in the assignment portfolio. The discovering of new competencies is highly facilitated by the socialization process (Nonaka, 1991; Nonaka and Takeuchi, 1995); that is, learning new tacit knowledge from the experts by observing their way of doing and receiving feedback from them when rehearsing. This step corresponds to the low to limited autonomy portion of the autonomy curve.

To choose the new development assignments, the colleague must tap in his interests, insights, intuitions, and hunches. Leveraging existing tacit competencies stored into the colleague's and company's sleeping base assignments can facilitate the selection process (Nonaka and Takeuchi, 1995).

Diagram 3.3.6

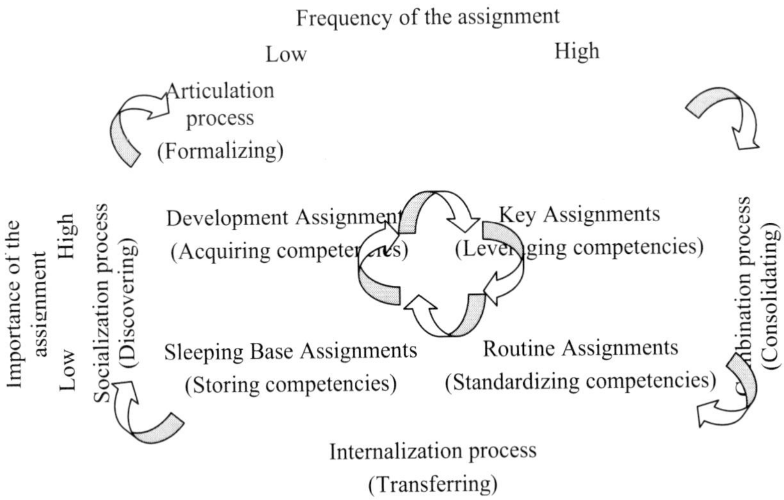

3.4 COOPERATIVE INNOVATION

The first three steps of the model of cooperative contextual change and the use of the protocols require the driver and those who have to implement the change:

- To check the applicability of the model and bring awareness of the need for change;

- To set the team and establish the conditions of collaboration and community to enhance the team's stability;

- To initiate and regulate actions, which ensure individuals' powerfulness to enhance team performance.

Once these conditions are in place, it is necessary to develop a solution-oriented mindset, which focuses on forecasting, and shift from a problem-solving orientation, which is backward looking. Once a solution orientation is in place, emphasis is placed on stimulating innovation and creativity, which enhances team competition (Choo and Bontis, 2002a).

Problem-solving focuses on finding the root cause of a problem as a preliminary condition to solve it. Trying to understand "Why did the problem happen?" results in locking people into reflecting on the specific problem within a past context, rather than exploring an understanding of the system and its future context (Watzlawick, Weakland, and Fisch, 1974).

Problem-finding is often offered as an alternative reflective approach. That asks the question, "What is the problem?" Focusing on "What is" instead of "Why did" challenges individuals' immediate interpretation of the perceived problem and its context, broadens the field of inquiry, reveals hidden dimensions, encourages "out-of-the-box" solutions, and leverages existing implicit or explicit, individual, or collective knowledge (Spender,

2004). However, a problem-orientation still exists, which requires waiting until the "something has gone wrong" has reached a level of awareness to trigger a response.

Encouraging the adoption of a forecasting/anticipation orientation means encouraging and facilitating individuals to engage new perspectives and ideas by leveraging their tacit knowledge, integrating their feelings and intuition to explore new contexts, and decision-making without asking for justification (de Bono, 1985). Solution orientation focuses on the question "How to make it work in the future?" (Watzlawick, Weakland, and Fisch, 1974). The process leverages all perspectives without mixing them (de Bono, 1985). Once a workable, future context is accepted, a solution orientation is used to provide an answer for achieving the desired goals. Solution orientation encourages individuals to engage in a process of learning to learn, which creates the condition for double-loop learning (Argyris, 1991). Within the "new" context, exploring the root causes of the existing "problem" is based on establishing learning conditions, which facilitate answering the question, "Why are we doing this?"

3.4.1 GEI—Groups for Exchange and Improvement

A group for exchange and improvement (GEI) is composed of people belonging to different disciplines, functional areas, and working on different projects. These individuals meet to find solutions to one or more issues and provide each other with suggestions for improvement. For GEI to be effective, a representative sample of participants is necessary to create the required variety in points of view. For a GEI to be effective, a minimum number of participants (nine) and of different topics or projects to be discussed (three) is necessary in order to create the required variety in the points of view. The optimal size of the groups for each subject is five to seven people, and no more than six subjects should be discussed.

1. In the first plenary session, the issues to be discussed are briefly presented and assigned to a sub-group. Each group is identified with a letter (e.g., from A to E) and assigned to one specific room (e.g., room one for group A to room five for group E).

2. Each sub-group discusses the specific issue using, for example, the six thinking hat technique, which every participant should have been trained on. The outcome of such discussion is a short written presentation on paper board sheets, where the group presents its solutions and/or actions plan (supported if applicable by data, documents, and/or information that are attached to the paper board presentation).

3. Next, the facilitator asks the teams to permutate. Each team visits the other team's write-out (team A goes in room two, team B goes in room three … team E goes to room one) with the goal of being exposed to others' work to challenge their perceptions and understandings. Each team member reviews the other team's presentation and support material. Using colored Post-Its, each individual writes his contribution using the six hats color code, and sticks it near to the point he wants to focus on. The facilitator tracks the analysis and calls time so that each team has reviewed and made a contribution within the allocated time.

4. Once each team has permutated and visited all other teams' works, teams return to the starting room, read

the Post-Its, and integrate the contributions into its own presentation.

5. Once the integration is completed, a plenary session is called for each team to present the final outcome of the GEI. Each individual must commit to the identified solution and the action plan.

A GEI allows different project teams to share their experiences and best practices, collect suggestions or ideas for further improvement, and discover opportunities for synergies among different teams. Each project team prepares a written presentation of the project, the issues faced, the competencies leveraged, and the solutions implemented. Because GEI is an open forum, diverse inputs from all interested parties are heard and discussed. This discussion encourages a shared understanding of the purpose that emerges among the participants (Jassawalla, Sashittal, and Hemant, 1998). A GEI is an innovation-supportive context because it is designed to allow people to share information, exchange and develop ideas, and make disagreements explicit while preserving the relationship.

It is structured as a ritual that help people to feel included, recognized, and supported (Schutz, 1958, 1966, 1982, 1984) to overcome us-versus-them thinking, reduce stress, challenge preconceived ideas, explore untested assumptions, expand boundaries, bridge unlinked concepts, and leverage difference to create value (de Bono, 1985; Jassawalla, Sashittal and Hemant, 1998).

Implementing a GEI ensures interaction between people, allows them to articulate and make explicit their tacit knowledge about individuals' areas of expertise, and creates the conditions for new, emerging explicit knowledge to be integrated in the process of innovation. This is possible because people can use others' knowledge, leverage others' creative insights, and identify synergies to set up innovative processes and products (Jassawalla, Sashittal, and Hemant, 1998).

3.4.2 Silent Roundtable Protocol

The silent roundtable protocol is the equivalent of a written brainstorming.

1. The facilitator announces the topic to be explored or discussed. Each participant writes down his own contribution in a short sentence but limits it to no more than three major points.

2. The facilitator invites each participant to read exactly what she/he wrote, starting with individuals who are usually silent. When somebody is reading, the facilitator writes the ideas on a paperboard. The other participants must keep silent, neither comments nor arguments are allowed during or after the reading. When a person has finished his reading, the facilitator invites another person to read, making sure to keep the more outspoken people last. If someone raises an idea already read, the facilitator will note it on the cluster. Once all ideas are presented, the facilitator interacts with the participants to organize the ideas into clusters. If the issue is critical and anonymity is necessary to enhance the expression of the participants, the facilitator can choose to collect the contributions and read them randomly.

3. Once all ideas are presented and put into clusters, discussion of the clusters focuses on singularity and interconnectedness.

The use of the silent roundtable protocol is useful because:

- It allows everybody to speak, especially the less vocal individuals, while reducing the risk of domination by the outspoken individuals;
- It helps to share ideas without criticizing or debating them in the moment by reducing judgmental attitudes and increasing the chance for unusual ideas to be listened;
- It requires individuals to think about what they want to say and prioritizes their contribution based on what is really important for them;
- It facilitates the organization of ideas by identifying themes, principles, processes, and/or outcomes.

The silent roundtable is less effective for leveraging others' ideas because it eliminates the interaction between participants. The six thinking hats technique (de Bono, 1985), discussed earlier, can be used to facilitate leveraging after the silent roundtable.

3.4.3 Call for Ideas Protocol

The call for ideas protocol is designed to facilitate the gathering of original solutions. It has three steps.

1. First, the idea client, who is facing an issue and seeking a solution, provides facts about the problem without pro viding any explanation and identifies possible causes and tentative solutions. Idea suppliers write their ideas and name on Post-Its (one idea per post-it). The Post-Its are collected by the idea client. This step should be kept short, and the ideas produced by the participants should not be discussed in the meeting.

2. After the meeting, the idea client analyzes the ideas produced and schedules separate meetings with the idea suppliers whose ideas he wants to deepen. In these one-to-one meetings, the idea supplier can freely raise, discuss, and test suggestions. He can bring a naïve or challenging approach to the issue discussed but needs to discuss positive instead of negative attributes. The idea client commits to listening to the arguments provided without debating them.

3. The idea client may develop workable solutions, organize new one-to-one meetings, or begin another round with new idea suppliers.

3.4.4 Reasons-or-Results (RorR) Protocol (Solution-Oriented Mindset versus Problem-Oriented Mindset)

The RorR protocol (Table 3.4.1) trains individuals to develop a solution-oriented mindset. Individuals are paired and asked to think of an issue that must be resolved (If the group is uneven, one of them observes a pair). In choosing the issue, individuals must feel comfortable enough to speak of it with their partner. People are given ten minutes to complete this phase.

1. Partner A, in two minutes, describes the issue to partner B, who acts as a consultant to help A to find a solution. Partner B questions A for no more than ten minutes using the questions presented in the R-Pf (see Table 3.4.1). Next, partner B uses the questions from set R-Sf (see Table 3.4.1) to question A. Partner B records the answers given to the two different sets of questions.

2. Partner A and partner B debrief the answers and the process and identify the pros and cons for each outcome and process. The team closes by discussing how to improve their day-to-day interactions.

Set R-Pf (Reason, Problem-finding)	Set R-Sf (Result, Solution-finding)
- What isn't working? - What is causing this problem? - What does it prevent you from doing? - Why is it serious? - How long has this problem existed? - Why has the problem continued? - Who is responsible for resolving this problem?	- Describe precisely the situation and the results you want to achieve. - What prevents you from achieving it today? - What do you control in resolving this problem? - Where, when and with who are you going to do it? - What are your current resources which will help you to succeed? - How will you know the task, plan, or project is complete? - What additional, positive consequences might be observed?

Table 3.4.1—Contrasting R-Pf (Reason, Problem-finding) approach to R-Sf (Result, Solution-finding)

PART FOUR: SUBSTANTIATING CASES OF THE IMPLEMENTATION OF THE M3C

The two integrative case studies reference the turnaround of two industrial companies as a means for discussing the outcomes obtained implementing the model of cooperative contextual change.

The first case is an interview with the industrial director of an SMS company specializing in manufacturing rubber products for the automotive and sports industries (let's call it "RC"). After having been trained to the M3C and the contextual drivership methods, the industrial director participated to the turnaround of the company and successfully drove the change process in the industrial department in the period January 2002—December 2003.

The second case is organized as a critical learning history of the successes and pitfalls occurred during the implementation of the M3C in the French subsidiary ("FS") of an American multinational company ("AMN"). This change management consulting and training mission was performed between March 2002 and June 2003.

The two cases are presented in a two column format (Kleiner and Roth, "How to Make Experience Your Company's Best Teacher," Harvard Business Review, Sep-Oct 1997).

- The right-hand column (and/or normal fonts) describes the facts concerning the company, the steps of the mission, and the tools that have been or will be put in place by the management as an outcome of the program;
- The left-hand column (and/or italics) describes the theoretical body of knowledge used as a reference to explain

the events and the outcomes and to design and implement the change processes or the consulting missions.

Using such a format proves to be useful for a further use of the cases as pedagogical tools. If the students are provided with a copy of the case where the left-hand column is blank, they can fill it in to show they understood the tools and master their use to explain a change process. When they are provided with the integral version of the case, they can autonomously confront their version with the teacher package, and therefore self-learn.

4.1 THE CHANGE PROCESS AT THE INDUSTRIAL DEPARTMENT IN THE RUBBER COMPANY

4.1.1 The Context of the Rubber Company

The Rubber Company (RC) is a private-owned SMS company specializing in the manufacturing of rubber products, working as a tier one subcontractor for automotive constructors (50% of RC's turnover) and sport products firms (50% of RC's turnover).

Headquartered in a twenty-thousand-inhabitant town in the middle of France, in December 2003, RC employed around seventy people in France and had a turnover in excess of 11 million Euros (60% the French operations, 40% Tunisian operations).

This case, based on several interviews with Virginie Papin, RC's industrial director, describes (right-hand column) and comments (left-hand column) the change process that she started in January 2002 to turn around the industrial activities department.

At that time, the crisis was the strike in the automotive industry, and RC's French factory was not profitable enough to be viable in the medium run.

The Tunisian subsidiary could be expanded to get all the production at much lower costs, while the other functions of the company (administration, sales, engineering) would have been kept in France to stay close to the clients. The decision of delocalizing to Tunisia the production of the low added value items would allow reduction by two-thirds of the French workforce. However, before making this disruptive decision, the CEO wanted to give a last chance to the French factory.

In November 2001, Ms. Papin started a six-month part-time training where she was exposed to updated management methods and techniques. During the program, she became acquainted with the model of cooperative contextual change and the drivership protocols.

In December 2001, the former technical director in charge of the products/processes design and engineering, the product development, the production, and the purchasing retired. Ms. Papin was then appointed as the new RC's industrial director after five years spent at RC as quality assurance manager.

Four months after Ms. Papin's promotion, the incumbent production manager resigned. Employed by RC for fifteen years, he expected to be appointed as the new industrial director and did not want to work for Ms. Papin. A major factor of possible conflict disappeared, easing the implementation of the change process.

No other major interpersonal conflicts existed in the industrial activities department. However, because of the rigid boundaries and ineffective interfaces between functional units, in the former hierarchical functional structure, people were not accustomed to cooperate, and the goals of each function loosely aligned with the global goals of the company.

4.1.2 Starting the Change Process at the Company Level

Question/Answer 1

How did the change process start at the Rubber Company?	
As recommended by the M3C, the CEO succeeded in mobilizing people's energies toward the protection of the stake by leveraging a triggering event, in this case the retirement of the technical director. In the contextual analysis phase concerning the CEO's level of responsibility, no cleansing process was necessary because the company was not affected by major conflicts between individuals or teams. In fact, what mattered was: - To make explicit what was at stake; - To design a sound organizational alignment map by gearing the stake down into specific objectives and purposes; - To link the organizational alignment map to the driver's added value matrix, to make sure that all the directors had correctly understood which were the stakes to be protected by each department and were able to gear them down into appropriate operational objectives for each departments' units.	In his first meeting with the senior management team, early January 2002, the CEO presented the results of an audit carried out by an external consultant. He announced that, in the next fifteen months, it would be necessary to increase the effectiveness of the communication and decision-making processes to protect the viability of RC's French factory. The CEO decided to modify the organizational structure of the company, shifting from a classical hierarchical functional structure to departments and units related to the activities of the value chain. He defined then specific goals for each department, asking each director to identify appropriate measurable performance indicators for each activity. As director of the industrial activities department, my primary goal was to increase the profitability of the production in the French factory. The purpose of my mission for the next fifteen months was to demonstrate that the French factory could reduce its production costs while maintaining the quality of the products.
The CEO succeeded in linking the Organizational Alignment Map to some simple and sound performance indicators, to provide everybody with a meaningful system of measurement of the effectiveness of the change process (Anderson and Ackerman Anderson, 2001).	I agreed with the CEO on the following performance indicators to measure the effectiveness of my action: - 100% on time product delivery, - growth of the production capability, of the machines occupation time, and of individuals' productivity - reduction of inventories for final products, raw materials and accessories.
Clearly identifying the constraints facilitated the understanding of the framework within which people can act, and therefore helped the directors to free up resources and to identify their own creative solutions (Watzlawick, Weakland and Fisch, 1974; Haley, 1973, 1976).	The reduction of the costs and the growth of productivity had to be achieved with no investments nor increase in direct or indirect staff. I was free to design the change strategy for the industrial activities department and to identify the operational indicators to follow up the process.

Question/Answer 2

<table>
<tr><td colspan="2">How did you use what you had learnt in the training to facilitate this process?</td></tr>
<tr><td>Being exposed to a systemic approach to change allowed Ms. Papin to foster the change process by creating the conditions for the implementation of a knowledge-in-action approach (Spender, 2004).

In fact, it helped her to build self-awareness, step back from the contingent events, and have a broad perspective of her and others' work (Anderson and Ackerman Anderson, 2001).

To foster knowledge-creation and facilitate change processes, a driver has to feel comfortable with the issues related to the distribution of power and the control of the activities and the fields of meaning, which are critical elements in any change process. This is necessary to correctly understand the importance of managing meaning and spreading effective practices (Spender, 2004).</td><td>Being trained to the model of cooperative contextual change and to the contextual drivership protocols helped me to understand the call for change, to reduce my resistance, and to improve my communication skills.

For example, I had to accept that the commercial relationships with the customers and with the suppliers, as well as the product/process design would have not been carried out by the industrial activities department.

This was particularly difficult for me because I had the impression that I was going to be excluded from the critical decision-making processes, and I would have lost the most interesting part of my job.

Later in the change process, I discovered that because of my former quality assurance manager position, I had a preconceived idea of what was the role of the industrial director and of the production activities.

I could then act as an agent of change to federate people around the process by using the tools I learned and making them learn these tools.</td></tr>
<tr><td colspan="2">Among the RC Directors, Ms. Papin was the only one who attended the training seminars on the model of cooperative contextual change.

This detail is particularly interesting to substantiate the model, because it shows that the contextual analysis and community norming phases can explain the processes and the behaviors of the CEO and his senior management team, even if they did not know the model of cooperative contextual change.</td></tr>
</table>

4.1.3 Starting the Change Process at the Industrial Department

Question/Answer 3

<table>
<tr><th colspan="2">How did you start the change process in the industrial activities department?</th></tr>
<tr>
<td>It might seem that Ms. Papin spent a long time on the contextual analysis phase. However, this has been one of the key factors of success of the effective implementation of the model of cooperative contextual change.

In fact, because she was an insider of the company and of the industrial activities department, she needed to step back from the system if she wanted to be able to reframe her vision of it.

Using the contextual drivership protocols helped Ms. Papin to have a distant, outsider view of close things (Ury, 1991).

By observing and listening carefully, she could leverage the common knowledge of the people working in it (Grant, 1996a, 1996b; Spender, 2004) and therefore understand the functioning of the system she was going to change.

It helped her to identify the elements she could leverage to create the conditions for a real second-order change. It gave her the legitimacy to trigger a second-order change for the people working in the industrial activities department. (Watzlawick, Weakland and Fisch, 1974).

The appropriate (and consistent during the time) use of the reliability mtrix and autonomy curve protocols gave Ms. Papin the elements to build a sound driver's added value matrix and competencies development cycle matrix for each unit manager.

As a consequence, she could see that a new position (the supply and production planning manager) was necessary to realign the functioning of the overall units of the department with her organizational alignment map.</td>
<td>I spent the first three months making an audit of the department. I wanted to understand why production was not profitable, and which factors could be leveraged for improvement.

I wanted to have a deep understanding of the organizational structure, the machines, the competencies of the people.

I used the reliability matrix to formalize all the tasks that the units' managers were carrying out, which were their competencies and their commitment.

I did not do it for the operators, and later I discovered that this had been an error. At that time I wrongly considered that the tasks of a machine operator were evident…

Once I had filled in the reliability matrixes for each of their activities, I assessed the perception I had of their level of autonomy.

In this way I could understand their strength and weaknesses, and identify their needs in terms of competencies development.

Having mapped the reliability of each unit manager in my early audit and having kept track of it proved to be very useful.

It allowed me to choose the right person to take my former position of quality assurance manger and to understand that I had to create the position of supply and production planning manager that did not exist before.

It allowed me to redesign the activities of each unit manager to make them focus on their core activities and managerial duties.

For example, I understood I could upgrade the activities of the technical supervisor to make him a real industrial development supervisor.</td>
</tr>
</table>

continued on next page

Correct understanding the systemic links between the elements of the system she was going to change helped Ms. Papin to optimally leverage all the available resources, and therefore to successfully protect the stake within the given constraints. By making reference to Senge's model (Senge et al., 1994), the use of the contextual drivership protocols since the beginning of the audit helped her - To develop her personal mastery and expand her personal capacity to create the results she most desired; - To modify her mental models and therefore increase the effectiveness of her actions and decisions; - To enhance her systems thinking and overcome the lack of an external view in carrying out the change process and implementing the model.	The mapping helped me to change my management and communication methods and to customize my management style to each person to make sound and appropriate decisions. Before, I was used to have the same management style with everybody, because I thought it was the way to be fair.

Question/Answer 4

How did you communicate the stake of your mission to your colleagues?	
Understanding the meaning of a goal is necessary to create knowledge and to develop best practices (Spender, 2004). The three months spent auditing the industrial activities department helped Ms. Papin to design a sound organizational alignment map to gear down his stake into detailed objectives for all her colleagues. Then, she leveraged it by adapting her messages to the needs and features of her colleagues at all levels. In this way, she introduced the first element of stability in the most appropriate way: without increasing the level of stress of her colleagues while outlining an action plan to implement an incremental but steady change in the department. Ms. Papin correctly decided to drive the change by setting the drivership relationship at a limited level of autonomy. In fact, she recognized that her team's level of reliability was quite low (mid-commitment, low-competencies) to protect a high stake (increase the profitability of the production). As recommended by the autonomy curve protocol, she identified several steps to be achieved to get the overall goal: first reduce the late-time production, then rethink the production process, then reduce the throughput production time and the indirect costs. She systematically controlled the achievement of each step before passing to the next one. Ms. Papin acted as a tutor, training her colleagues to answer their need of support. While she firmly kept focused on each step of her action plan ("You do not have the choice, we have to do that…"), she allowed her colleagues to go at their pace ("…but you have the choice on how achieving it. Let's take as much time as necessary). The goal is not negotiable, but the way to achieve it is negotiable.	If I had told immediately and explicitly the stakes of the company and of the industrial department, they would have been scared. Thanks to the audit, I could identify one meaningful goal on which everybody could agree: reducing the throughput production time from eight weeks to six weeks. In this way I could make the stake explicit without directly telling it. In fact, if we wanted to achieve it, we had to change the organization. By reorganizing the department, we could increase the productivity, and therefore the profitability of the production. I did not announce it to all the industrial department staff but just to the managers of the units. I told them that the only thing that mattered was the production, and what they had to do first was to reduce the late-time production to zero. This was the only goal they had to give to their team leaders and operators. I did not want to stress people with a too-ambitious goal. They were operating with 20% late time on eight weeks throughput production time; if I had asked them to reduce the throughput production time to six weeks, and to be on time, it would have been too much for them. I had to go step by step. The first step was to become able to produce on time, regardless of the throughput production time. The production manager told me that to produce on time would have required at least six months. I told him that on-time production was a global aim of the restructuring, and any attempt to reduce the throughput production time without achieving first the on-time production would have been nonsense. Therefore, we would have taken as much time as necessary, and I would have supported them in their efforts by providing them with the means they needed to achieve this goal. I then met each unit manager and agreed with them on specific units goals, each with a clearly identified purpose. Making explicit the purpose of each goal helped them to find the meaning of the actions I was asking them to accomplish and understand the consistency of the change process.

continued on next page

	I spent a lot of time teaching them the methods I had learned in the training, and I allowed them the right to make mistakes. It helped them to reduce their level of stress and therefore focus their resources on acting instead on resisting.
Ms. Papin understood that to make her learning effective she had to bridge it with her acting in the firm (Spender, 2004). By teaching her colleagues, she contributed to sharing it and embeding it into new common practices. This helped people to start behaving differently, and therefore to reshape their perception of the reality, contributing then to the creation of a new identity for the industrial activities department (Skinner, 2001). Allowing the right to make mistakes, and using them to learn, is a condition to reduce resistance to change and facilitate competencies growth (Senge et al., 1994).	

4.1.4 Implementing the M3C at the Industrial Department

Question/Answer 5

What happened then?	
Achieving the first step took around six months, that is 40% of the global time (fifteen months) allowed by the CEO to accomplish the change process. But this achievement was important because it contributed to increase people's commitment and competencies, allowing Ms. Papin to modify her drivership style by setting the relationship to a sufficient level of autonomy. She could then announce to her colleagues a far more ambitious goal, without increasing the level of the stress in the department.	In September 2002, the production was on time, and we had reorganized the supply system. In December 2002, I announced that the new goal was then to increase the profitability of the production and that we had three months to do what was necessary to show to the CEO that we could get it.

Question/Answer 6

How did you make sure that your managers would have adapted their behaviors to the new structure of the industrial department?	
The audit and the identification of the steps to protect the stake allowed Ms. Papin to carry out the contextual analysis phase and to install the first element of stability. To install the second element of stability, Ms. Papin chose to carry out the community norming phase in the same way the CEO did. However, since the level of autonomy of her colleagues was limited, she had to closely drive this process, adapting her drivership style to the different behaviors of her colleagues. She was more direct with those showing lower reliability, and just oriented those showing higher reliability. In this way she could preserve the commitment of the change agent, increase the commitment of the obstacles, watch out for the opponents, and provide the opportunists with convincing results.	Here, again, I went step by step. First, I asked each manager to present to the others his job, his objectives, the activities he carried out, the information he could provide to the others, and the information he needed from them. Then I asked them to meet and decide their rules of functioning and the way they wanted to handle the interfaces between their units. The only imperative I gave them was that the production had never to stop. I left them free to choose the way they wanted to interact. The individuals that were ready for a greater level of autonomy were pleased, because they understood they could have more fun and freedom in their job. Those that were more execution oriented were disturbed by my determination to increase their level of autonomy in our relationship, because it implied that they had to take more initiative and therefore work harder. Sometimes I had to impose my point of view to those people that were more reluctant to buy into this co-building and co-opting process.

continued on next page

	Second, I made clear to everybody that I wanted them to stop waiting for me to provide them with the information. I explained to them where and how they could get it, so that they could look by themselves for it.
	I taught them how to use the CAPS—Computer Aided Production System of the company, which had been installed the year before but nobody was using because of lack of training and no constraint to do it.
Ms. Papin appropriately leveraged the slots of active support protocol to increase the level of autonomy of her relationship with her colleagues.	I refused to answer their questions concerning information they could find on the CAPS, and therefore I forced them to use it and to develop their own information network, internally and externally to the industrial department.
By making reference to the competencies development cycle, she identified the use of the CAPS as a development assignment, which would eventually helped them to reshape the content of their key assignments and therefore their core activities in the driver's added value matrix.	Today, they come to me to ask for information only when they really could not find it.
	Third, I taught them the methods I use to make a decision, and I made clear the limits within which they could decide alone.
By retaining information on purpose, Ms. Papin used the technique of "benevolent sabotage" (Watzlawick, Weakland, and Fisch, 1974) to get around resistance to change.	With regard to the activities that they could carry out and decide autonomously, my main role would have been to enforce the rules of functioning and the interfaces they had co-built and co-opted, to support them when necessary, and to arbitrate when they would have not been able to find an agreement.
By promoting the use of the IT system, she contributed to leverage RC's existing data and information and to make it a critical factor of improvement in production.	With regard to conflict handling, I asked people to find together a viable solution. If they cannot solve it, I arbitrated. This is the way I still act today.
	For example, I had the case of a conflict between a new employee who had a diploma but little experience and an old one who had a lot of experience but no qualification.
However, the appropriate use of IT is not enough to provide significant strategic advantage if the relationships between people and the issues related to the distribution of power in decision making are not addressed (Spender, 2004).	Since they were unable to find an agreement alone, I arbitrated in favor of the old one, because he was showing a greater commitment to the company.
	By this time, the units' managers have geared these methods down to their team leaders and operators.
To do so, Ms. Papin put into practice the basic of drivership, that is the separation between the roles of hierarchy, leadership, and management.	In the past, the production manager was expected to give the order of using one machine or another.

continued on next page

By teaching her unit managers how to make decisions, she gave them the possibility to initiate and regulate actions to protect the Stake. However, first she established ground rules to handle the hierarchical relationships, keeping for her the arbitration responsibility. Since the beginning of the change process, she could then install the forth element of stability by making her colleagues understanding the importance of a solution-oriented mindset. As recommended by the drivership protocol, when reliability is unstable, it is important to focus on commitment increase before competencies growth in order to start a virtuous circle. In fact, commitment makes reference to the "decision to participate," and implies that the colleague is willing to accept the field of meaning of the company, and to contribute to shape it when required (Spender, 2004). Appropriately leveraging the autonomy curve protocol helped Ms. Papin to drive the increase of reliability of all her colleagues in the industrial activities department. In fact, she could make their competencies development cycle moving: information gathering (IG) and decision making (DM) became first a development assignment, then a key assignment. When the unit managers became comfortable with this new way of working, Ms. Papin asked them to move IG and DM to routine assignments. The ultimate result was a modification of the unit managers' driver's added value matrix: some of the unit managers' responsibilities shifted from core activities to subsidiary activities. This allowed the unit managers to concentrate on more important activities at a higher added value. At the same time, it activated the competencies development cycle of the team leaders and the operators, by providing them with new development assignments. The outcome is then an overall improvement of the autonomy of the people working into the industrial activities department and an increase of its performance on the long run.	If a machine was out of order and the production manager was not available, the team leader could not decide alone, and the production was stopped. Today, the team leaders have learned the methods to decide which machines should be used, and they can make on-the-spot the most appropriate decision according to the context they face.

Question/Answer 7

Then, how did you implement the change process?	
By making reference to the autonomy curve protocol, Ms. Papin correctly set regular fixed points, which helped the unit managers' to complete the community norming phase. Once the unit managers became able to handle their interfaces at a sufficient level of autonomy, Ms. Papin could modify her driver's added value matrix, shifting the regulation of unit managers' interactions from a core activity to a subsidiary activity. Accordingly to the autonomy curve protocol, she just went on carrying out fixed points to have a focused control on the key factors of success of unit managers' work. Once the second element of stability was installed into the system, Ms. Papin could start the contextual drivership phase. Training people on the use of the reliability matrix and the autonomy curve protocols helped Ms. Papin and her team to adopt systemic relationships methods to initiate and regulate actions (Hersey and Blanchard, 1972). These protocols became a common meta-language, which increased their capability to reframe the perception each person has of the context, and focus on solutions (Watzlawick, Helmick Beavin, and Jackson, 1967; Haley, 1973, 1976, Spender, 2004).	To install such state of mind, at the beginning I met with the managers and the team leaders every Monday morning to make them exchange on what happened last week and what they were going to do in the coming week. My role was to facilitate and mediate. Once they got the habit of communicating when needed, these meetings were no longer necessary, and nowadays I just carry out a monthly meeting with each of them. Then I taught the managers and the team leaders how to use the reliability matrix and the autonomy curve. Nowadays, before every monthly fixed point they plot their perceptions, I plot mine, then we confront them. The plottings are usually very similar, and it helps them to better accept the way I manage them. When the plottings are divergent, it helps both of us to step back and reconsider the relationship we have. We discuss then on how to improve it, and this helps me to fine-tune my managerial style.

4.1.5 Leveraging the Contextual Drivership Method

Question/Answer 8

What has been the impact of using the autonomy curve and reliability matrix protocols?	
The use of the drivership protocols helped Ms. Papin to have a better understanding of the global functioning of the system and identify the factors that she could leverage to improve the use of the knowledge existing in the company.	At the beginning, when I was asking a colleague to do something new, he often told me that he did not want to do it because he was not motivated. Using the reliability matrix and the autonomy curve has been very useful to make them become aware that the blocking factor was often a lack of competence. In most of the cases, providing them with appropriate training has been enough to make them overcome their resistance.
However, providing training to increase competencies would not have been enough, if Ms. Papin did not provide her colleagues with - The purpose of action necessary to create meaning, - The opportunities to put such learning into practice (Spender, 2004).	As a consequence, they felt more confident and motivated, their adaptability increased, and so did their performance.
These indicators have some interesting features (Spender, 2004; Grant, 2001, Polany, 1962): - They are all grounded on the actual activities performed by the operators and are connected to their fields of meaning and set of practices; - They measure proficient performance, but at the same time they keep into account the role and impact of tacit knowledge. In fact, while some are quantitative and focused on production output (punctuality, quality, productivity), others are more qualitative and focused on those behaviors which can make the community a place of knowledge creation and sharing (respect of rules, relationships with the others, self-organization).	To assess the performance of the machine operators, the production manager and I built an assessment grid that makes reference to the five elements that we consider can be mastered by the operator: - Punctuality and respect of rules, - The quality of work, - The productivity of work, - The relationships with the others, - Their capability to self-organize and self-manage their work.
The extremely high rate of consistency of the rating demonstrates a great sincerity of Ms. Papin's colleagues and their good understanding of the purpose of the exercise.	We assessed all the operators on this five elements grid, then we met them individually and we asked them to rate themselves. Ninety percent of the operators rated themselves exactly as we did, with no difference. We asked them to tell us their expectations and give suggestions to better use their competencies. We could then together identify their strengths and weaknesses and how to leverage the former and improve the latter.

continued on next page

Providing people with elements of measurement of their performance related to their day-to-day practices contributes to provide meaning for an effective knowledge integration, and therefore to enhance their commitment to participate in the change process (Spender, 2004).	Today, each managers carries out monthly fixed points with each of his team leaders and each team leader with his operators. I plan to train them in the drivership protocols and other appropriate management techniques to increase their reliability and therefore their level of autonomy in the relationships with me and their colleagues. Filling in the autonomy curve, the reliability matrix and the five elements grid allowed us to modify the salary grid to redefine the job descriptions, allocate people according to their competencies and their expectations, and therefore increase the versatility of our workforce. The new salary grid is explicitly related to the level of autonomy that the individual is able to establish with his colleagues. This helps to reduce the conflicts due to the perception of unfair treatment and pushes people to try harder.

Question/Answer 9

How did your colleagues contribute to the adjustment of the change process?	
This example shows an interesting practice that Ms. Papin managed to foster the use of the feedback protocol with the objective of creating a double-loop learning (Argyris, 1991). This practice puts Ms. Papin's colleagues in a double-bind situation (Watzlawick, Weakland and Fisch, 1974) and is completely consistent with the autonomy curve and the reliability matrix protocols. In fact, if the colleague appropriately reports to her, he shows that he is ready for a higher level of autonomy in the relationship with her, and therefore he deserves a reward. If the colleague does not report in a proper way, he shows that he is not ready for a high level of autonomy, and therefore he does not deserve the full reward for the job he has accomplished. The final aim of this "game" between Ms. Papin and her colleagues is to support a mindset focused on continuous improvement and best practices formalization as a means to successfully implement the turnaround of the industrial activities department and create the conditions for a long-lasting increase in competencies and knowledge sharing.	At the beginning it was very difficult for them to provide me with appropriate feedback on the activities they were carrying out. They were used to coming and seeing me when they were facing difficulties, but they could not understand the usefulness of reporting their results, following up their actions, and telling me what had been done correctly and what could be done differently. To help them overcome this difficulty I decided to link 20% of their performance bonus to their capability of providing feedback. Even if they had correctly achieved their objective, to get the total bonus, they had to come discuss with me and defend their achievements, explaining to me what they did to get it. If they were not proactive in reporting to me, they had just 80% of the bonus, regardless of how good the results were. This forced them to take the habit of organizing a monthly meeting to present to me their activities. By that time I taught them how to formalize their performance, as well as their difficulties and their needs. I taught them the method to build an action plan with relevant indicators to successfully manage small projects aligned with the global project of the company and of the industrial direction.

4.1.6 Outcomes of the Implementation of the M3C in the Industrial Department

Question/Answer 10

<table>
<tr><td colspan="2">What main improvements do you observe in the functioning and the performance of the industrial activities department?</td></tr>
<tr><td>By making reference to Spender's different types of organizational knowledge (1996), it is possible to show that the way Ms. Papin carried out the change process facilitated the creation of a community of practices.

In fact, teaching her colleagues the techniques she had learnt during the training helped her to transform this conscious knowledge (individual, explicit) into an automatic one (individual, implicit) and an objectified one (social, explicit).

Then, by encouraging (and sometimes forcing) her colleagues to embed it in their daily practices she made possible the creation of collective knowledge (social, implicit), which finally became the automatic knowledge (individual, implicit) of the people working in the industrial activities department.

The creation of such community of practices made possible the introduction of changes that otherwise would have been rejected, as the link between salary and autonomy.</td><td>In the previous hierarchical functional structure, people lacked vision and were unable to anticipate issues.

Today, I provide my colleagues with all the important information concerning the medium term vision, and I ask them to take the actions necessary to anticipate what they believe could become an issue.

At the beginning, they informed me before taking any initiative; today, they know within which boundaries they can act autonomously; they come see me only when they have a doubt or they face an obstacle and need support or advice.

Most of the time they find their own solutions because they have understood that they are interdependent, and that they have to autonomously exchange information to make everybody work correctly.

For example, to develop the new products designed by the technical department, the product development manager interacts since the beginning with the other departments and with the quality assurance manager and the production manager. This was not happening before, unless the boss had decided so.</td></tr>
</table>

continued on next page

Blaming others is one of the most current ways people have to divert their energies from finding solutions to problems and therefore reinforce the problem (Watzlawick, Weakland and Fisch, 1974). By forbidding it, enforcing the rule, and showing the example, Ms. Papin forced her colleagues to use their resources to design and implement actions aimed to integrate different factors of production, as well as different levels of meaning with the purpose of effectively managing knowledge for the benefit of the community as a whole (Spender, 2004).	Another habit that has completely evolved is blaming somebody else. I have tackled this by establishing a very strict rule: blaming others is simply forbidden! When a problem arises, everybody has first to look for a solution that could be acceptable by the others, then discuss together to implement the most appropriate actions to protect the stake. Today, since everybody knows that the stake is the profitability of the production, everybody takes the initiative, when necessary, to make sure things go in the right direction. For example, when non-conformity in production is above 20%, the team leader (or the operator) stops the production and looks for a solution, instead of just waiting for the maintenance team to fix the problem. However, this solution-oriented mindset is limited to a curative action. They still did not have the reflex to preempt problems or to improve the production. Often they make confusion between solution and means. They think that to do better it is necessary to have more means and resources. My role today is to train them to observe and to have a reflective thinking approach and to find, whenever possible, smart solutions without additional means. For example, when they came saying that they had too much technical scrap, they provided me with many explanations to that but did not think of a solution. I then asked them to think what they could do to reduce the technical scrap.

continued on next page

At any time people face uncertainty, their capability to create new practices by leveraging their knowledge is reduced. It is then necessary for the driver to help his colleagues to overcome: - Ignorance, by providing them with the relevant data they could not acquire alone; - Indeterminacy, by providing them with the tools necessary to imagine scenarios that they will be able to implement and control alone; - Incommensurability, by using power and arbitrating when they face divergent issues that they cannot integrate into their mental framework (Spender, 2004). Ms. Papin completely implemented the first three steps of the M3C: contextual analysis, community norming, contextual drivership. With regard to the fourth phase, cooperative innovation, people adopt a solution-oriented mindset when they have to react to an issue or a problem. The next step is to create the conditions for a proactive use of such solution-oriented mindset. This is necessary to create the conditions for the next change process to be implemented.	They started thinking on this new goal, but they forgot the others. So my role is constantly reminding them the set of goals they have to achieve and reorient their efforts when necessary.

Question/Answer 11

What is the climate in your department today?	
Ms. Papin resisted the temptation of implementing a coercive or break change driven by fear and crises. On the contrary, she preferred leveraging accountability, fostering credibility, and supporting authenticity. Thanks to her steadfast discipline, constant direction, and simple implementation, she could successfully build tangible evidence that her plans make sense and deliver results (Collins, 2001). Her decision to make reference to the model of cooperative contextual change and implement it on an incremental basis paid off in the long run because it helped her to install the flying wheel effect, which will help her to face the next changes (Collins, 2001).	Stability increased, but it is still fragile. In 2002, people were convinced that the CEO was going to delocalize the production in Tunisia, and that even if they were given a last chance, they could not succeed. Today, they start believing that they can win the challenge, but everybody is waiting for the 2003 results to have the demonstration that their efforts paid off, that the method works, and that things will continue going in the right direction. They noticed that the company is investing again in the French factory: the shop floor has been repainted, a canteen has been created, new machines have been installed, wages increased. Thanks to that they are recovering trust and self-confidence. In the future I'm going to ask people to fill in a questionnaire to measure their satisfaction and create an indicator of the climate.

Question/Answer 12

What are the results of the implementation of the change model and of the contextual drivership methods?
Table 4.1.1 provides a comparison between the forecasted and actual budget for the fiscal year 2003, Table 4.1.2 the results for the industrial activities department, Table 4.1.3 a longitudinal overview of the main figures concerning RC's overall activities. These tables demonstrate the dramatic increase in RC's performance, which is mainly due to the turnaround of the industrial department. Ms. Papin thinks that people start to be proud of working for RC, and this positive climate is supported by an increase in their wages/bonuses and in their ranking in the job classification. While productivity and quality increase, fixed costs diminish, which allows an increase in the net profit after taxes and in the self-financing cash flow, despite the stagnation of the turnover. The increase in variable costs compared to the forecasted budget is due to the more intense activity, and it is largely compensated by the increase in productivity. According to her, the use of the cooperative contextual change model helped her to successfully combine theory E (economic and operational output) and theory O (organizational structure and climate) (Beer and Nohria, 2000) to overcome the forecasted performance and to create the basis for stable dynamic change in the future.

(Unit = k€)	Budget 2003	Actual 2003	Difference actual 2003 — forecasted 2003
Total turnover	6664	6476	-198 (-2.5%)
Self-financing cash-flow	3563	3747	+184 (+5.2%)
Value added	2856	2964	108 (+3.8%)
Operating profit	593	665	72 (+12.2%)
Energy	120	132	-12 (-10.0%)
Insurance	55	70	+15 (+27.3%)
Cleaning	160	195	+35 (+21.9%)
Wages Tenured Employees	1733	1731	-
Wages Part-time Employees	390	432	+42 (+10.8%)
Taxes	140	164	+24 (+17.1%)

Table 4.1.1 - Comparison between forecasted and actual budget for fiscal year 2003 (French operations)

Item	Jan 2002	Sep 2003	Remarks
Average production time	8 weeks	4.5 weeks	+45% improvement
Average late items	20%	7%	Despite the 7% late items, because of the 4.5 weeks' production time, all products are delivered on time.
Number of out-of-shelves in raw materials	2 per 7 days	1 per 45 days	Sometimes the raw materials are forwarded to the Tunisian subsidiary without diligent replacement.
Average purchased materials inventories	135k€	100k€	+25% improvement 2002 figures take into account only the raw materials. 2003 figures all purchased materials.
Average finished good inventories	240k€ 60 days	170k€ 30 days	+30% financial improvement + 50% time improvement
Increase in production capability (with no investments)		+10%	Introduction of background maintenance (no stop in production).
Hourly gross production added value per operator	37€/h	54€/h	+46%
Work time shifts	4x7h	4x6h	
% of transports by FedEx	10%	6%	The remaining express transportation is due to specific customers' requirements and billed on top.
Wages			+25% increase in operators salaries Forty-five people over seventy had an increase in salary. Change in the salary grid in relationship to the autonomy of each individual for a more fair and appropriate treatment.

Table 4.1.2 - Results for RC industrial activities department

(Unit = k€)	2000	2001	2002	2003	Difference 2003-2002
Total turnover	6006	6229	6162	6476	+314 (+5.1%)
Net profit after taxes and its percentage on turnover	86 1.43%	53 0.9%	-16 -0.2%	214 3.3%	+230
Self-financing cash flow	287	271	143	416	+273 (+290%)
Wages tenured employees Wages part-time employees Total wages	1433 396 1829	1795 404 2199	1702 439 2141	1731 432 2163	+29 (+1.7%) -7 (-1.6%) +22 (+1.0%)
Profit-sharing with employees and its percentage on net profit after taxes	6,5 7.6%	3 5.6%	0	27 12.6%	+27
Tenured employees Part-time employees Total employees	58 13 71	74 13 87	67 15 82	61 15 76	-6 (-9.0%) - -6 (-7.3%)
Average wage/tenured employee Average wage/part-time employee Average wage/employee	24.7 30.4 25.8	24.3 31.1 25.3	25.4 29.3 26.1	28.4 28.8 28.5	+3 (+11.8%) -0,5 (-1.7%) +2.4 (+9.2%)
Turnover/employee	84.6	71.6	75.1	85.2	+10.1 (+13.4%)
Net profit/employee	1.2	0.6	-0.2	2.8	+3.0
Profit sharing/employee	0.092	0.035	0	0.355	+0.355

Table 4.1.3 - Results for overall RC French operations

4.2 THE FRENCH SUBSIDIARY OF AN AMERICAN MULTI-NATIONAL COMPANY

4.2.1 The Context of the American Multi-National Company (AMN) and Its French Subsidiary (FS) at March 2002

FS is the French subsidiary of AMN, an American multi-national company, organized in several divisions, each of them dedicated to a specific type of industry. Each division is organized in three business units, each managing the companies located in their geographical area (North America, Europe, Far East). Each business unit manages several companies like FS, each specializing in one or two special products.

This case is based on the facts and information collected by the author during a six-month consulting mission, and provided by the CEO and the other managers of the company during the preliminary interviews, the training meetings, and the coaching sessions.

AMN is a streamlined company with few levels of operational management but with many functional directors, which creates a sort of matrix organization. Sales people do not respond to the directors of each company but to the sales director of each division. AMN and its subsidiaries' context is characterized by the conflicts arising because of the divergence of interests between salespeople, who sell worldwide and must focus on maximizing the volume of the sales, and production people, who produce locally and must focus on optimizing production (reduce costs and inventories) in order to produce as much cash flow as possible. Moreover, since AMN and EBU managers are used to interfere in subsidiaries' internal and operational activities, informal information bypass and decision-making shortcuts jeopardize subsidiaries' managerial stability.

FS belongs to the European business unit (let's call it EBU) of the packaging division (let's call it PD). FS produces one hundred million boxes and one hundred thirty million accessories for packaging food products (powders and biscuits, 80% of FS production and turn over) and specialty mass consumption items (cigars, perfumes, pens, watches, etc., 20% of FS production and turn over).

FS is located in a remote, economically depressed countryside location in the east of France, and it is the major employer (350 people, 270 long-term employees, 70 interim) in an area of ten thousand inhabitants. FS's turnover is 40 million Euros.

FS clients are French, European, American, and Asian. They are very demanding concerning the hygiene (for the food boxes) and the quality of the boxes (printing, assembling of the various accessories). FS's clients in the food industry cannot provide more than a two-month forecast of their demand for FS products, since their production varies according to the market, which is very unstable and competitive. FS's clients in the mass consumption market ask FS to produce a small series of special boxes requiring specific design and adaptation of the production tools.

AMN bought FS, a former very specialized small French competitor, in 1995, during a multimillion dollar's external growth project.

In 2001, FS equipped its factory with a brand new special six-color printing machine. At the leading edge of the industrial printing technology, this machine must be run by highly skilled workers and requires high level of production to be profitable. Any mistake in the setting or in the scheduling of the machine provokes low-quality printing and clients' rejection of the products; poor printing planning provokes shorter than optimal production series and therefore an increase in the production costs, due to the high setting time of the machine.

At the end of 2001, AMN faced a cash flow crisis due to:

- The high level of the debts contracted to buy competitors and to invest in new technologies;
- The reduction of the revenues because of clients' pressure on suppliers' prices.

Between 1995 and 2001, EBU appointed five different CEOs in the attempt to improve FS's performance. The first, the second, and the fourth CEOs were fired before the end of their first year, the third joined EBU as European technical director after twenty-four months as CEO of FS. The fifth CEO, a fifty-two-year-old Frenchman (let's call him CEO), has been working in the last twenty-five years as EBU financial controller (he was in charge of FS) and was appointed in September 2001 with the mission of:

- Increasing FS's cash flow to contribute to the reduction of AMN debt;
- Improving FS's performance in order to facilitate its selling; necessary to reduce AMN debt.

Since the beginning of his directorship, CEO knew he could not forecast how long he would have resisted at his place because of the hiring-firing policy of the company and because he was not expert in boxes printing and production. He had never directed a company. FS is organized as a typical staff and line organization (Diagram 4.2.1):

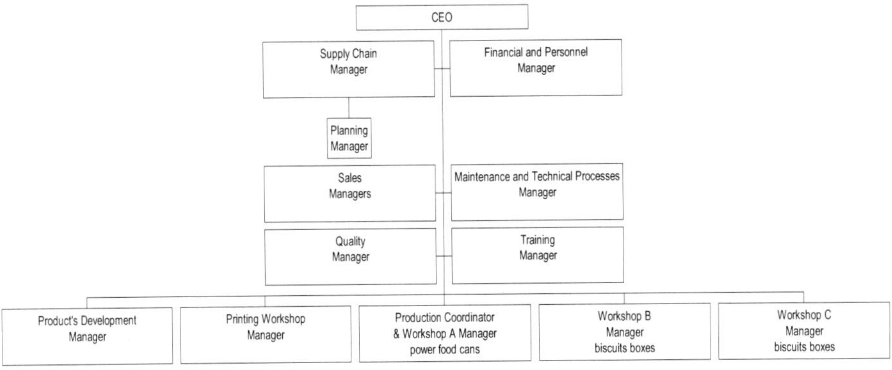

Diagram 4.2.1—FS Organizational Chart

Sales managers are hierarchically dependent from the PD's sales director, and functionally dependent from CEO.

Most of the managers are thirty- to thirty-five-year-old engineers with little senior management experience and coming from operations; the others are self-made people, with no formal higher education.

The relationships between these two populations are very conflicting, regardless the age:

- Young managers are quite unmotivated because of the remote location of the company, the difficult financial situation of AMN (no wage increase, no bonuses or benefits, no long-term investments in new equipments), the uncertainty concerning the future of FS;
- Old managers are resigned to see the company sold for the third time in ten years; that means lay-off and pre-retirement plans.

In March 2002 FS was facing three main operational problems:

- No coordination between salespeople and production people, with a special concern for planning and quality specification, with many last-minute changes that provoke chaotic and/or wrong production;
- High level of inventories for raw materials, low capability of organizing long production series to cut costs because of lack of selling forecast and of warehouses to stock products;
- Lack of discipline and of homogeneous managerial methods at all levels, due to the frequent change of CEO, which prevent people to coordinate their actions and to exchange best practices.

4.2.2 Implementing the M3C for FS's Management Committee

Table A

Real learning organizations are very skilled at making survival anxiety bigger than learning anxiety. Such a result can be achieved either by threatening people to increase survival anxiety, or by creating a safer environment for unlearning and new learning to decrease learning anxiety. The first way is the easier in the short term, but its by-product is an increase of resistance to change. The second way is the most effective on the long term, and the process can be greatly facilitated by good training, coaching, group support, feedback, and positive incentives (Schein, 2002).	In March 2002, CEO's primary objective was to ensure the increase of FS's performance in order to safeguard the workplace of the 350 employees directly working at the FS factory. The purposes of the implementation of the model of cooperative contextual change were: - To consolidate CEO's legitimacy in directing the company's turnaround; - To train FS personnel to leverage their specific skills and to adopt new cooperative managerial practices; - To make FS an example of an evolutionary company within EBU and ensure its long-term survival, either within AMN or within another group, in the event FS were sold.
Any change management program must therefore be initiated and supported by an appropriate training to force intensive early development and provide constantly-increasing professional *challenges (Quinn, Anderson and Finkelstein, 1996).*	The design of the project had to be tailor made and customized to create the most favorable conditions to ensure that the seven managers reporting to CEO (plus their five assistant managers and the three sales managers) could work more effectively together in order to start a virtuous turnaround cycle.

Table B

Commitment is fundamental to build a learning organization able to leverage knowledge, since it provides employees with a sense of identity with the enterprise and its mission (Nonaka, 1991, Salancik, 1977a, 1977b). Carrying out the qualitative audit allowed the author to: - Have a global understanding of FS's context; - Collect information on what issues each individual was facing and what their specific training needs and suggestions were; - Identify the main interpersonal, intra-organizational, and inter-organizational conflicts that were creating resistance to cooperation and information sharing and could jeopardize the implementation of the cooperative contextual change project; - Design an appropriate set of indicators useful to build a questionnaire to be administered regularly in order to measure the initial gap from the target and the progress made during the project.	The author wanted to verify the robustness of the commitment of the CEO to support the complete implementation of the project. He asked the CEO (who agreed) for permission to make a short qualitative audit. He carried out then with each manager concerned by the project an individual ninety-minute confidential interview. After the audit, the authors met again with the CEO to validate the training goals and process and to co-build the training program. Then the stake, the purpose, the goals of each phases of the training, as well the schedule and the topics were announced to all the managers concerned by the training program.

Table C

A successful cooperative contextual change program must have solid foundations. Setting and managing a change program consists of guiding and integrating the autonomous but interconnected work of highly skilled people (Argyris, 1991) who are able to find the right balance between competition and cooperation in order to create the conditions for co-competition (Nalebuff and Brandenburger, 1996). This climate of positive emulation is the necessary condition aimed to set up an organization skilled at creating, acquiring, and transferring knowledge, and at modifying its behaviors to reflect new knowledge and insights (Garvin, 1993).	One week prior to the kick-off seminar, the author acted as mediator to facilitate one-to-one meetings aimed to work on the major interpersonal conflicts existing between the participants. The aim was to "clean" the emotional backgrounds in order to: - Foster people's openness to the messages and the techniques that would have been delivered during the seminars; - Create the conditions for effective cooperation by separating interpersonal issues from professional issues.
Managers should focus their early action on facilitating individual and collective learning by creating a receptive environment, building trust, training people at teamwork, fostering collective reflection, and allowing public discussion (Garvin, 1993; Kleiner and Roth, 1997). Ground rule setting is necessary facilitate teamwork, discipline information sharing, and implement creative processes (Quinn, Anderson and Finkelstein, 1996). The result of this two-day seminar was an increase in team's cohesiveness by having all people committed to work all together to protect the stake. Of course, some of the participants were more resistant than others to such a surprising approach. However, since they had been identified during the audit, it was possible to manage them for the best, by leveraging as well the help of those participants already open to change. At the end of the community norming seminar, people were ready to question their individual and collective way of managing and learning and could pass to the second module, contextual drivership.	The training program started with a two-day community norming seminar (June 2002) aimed to: - Establish ground rules of conduct that the participant had to follow during all the training cycle. Co-opting and regularly using such regulation and self-regulation tools not only facilitated teamwork during the seminars but also helped the participants to internalize them. They started naturally to use them in their day-to-day work; - Create the conditions that will bring people to make their individual and collective commitments explicit, that is necessary to ensure long-term support to action plans; - Train people to the emotional leveling communication Protocol, a practical and easy to use way to give feedback by separating people from problems and manage conflicts in order to accept, prize, and promote diversity in the approaches to value creation; - Establish a consensual commitment process to facilitate action plans setting after decision-making processes to avoid implementation failures and preserve organizational alignment.

Table D

Training people to develop their leadership side is necessary to orient the knowledge creating activities of employees, to provide a conceptual framework that helps employees make sense of their own experience. It helps people to answers questions like “What should be?” “What are we trying to learn?” “What do we need to know?” “Where should we be going?” and “Who are we?” (Nonaka, 1991) Developing the management side is necessary to help learning facilitators to guide and integrate the autonomous but interconnected work of highly skilled people. It helps people to answer questions like “What must be learned and done?” “How should we learn and do it?” and “How can we do it better?” (Argyris, 1991)	The contextual drivership four-day seminars (a one-day seminar each two weeks in September-October 2002) were aimed to: - Fine tune system’s functioning interfaces’ quality and efforts’ consistency; - Acquire and co-opt leadership and management tools to be used by all members; - Make driver/colleague relationship rules explicit; - Facilitate appropriate feedback at any level to foster people’s development.
Ensuring the efficient and effective management of the interfaces helps to optimize people’s performance, improve systems’ stability, and verify organizational structures’ soundness. Thanks to the collective learning experience and the practical application of the tools during the seminars to actual company’s issues, the young managers acquired new insights on what happened and was happening at FS, EBU, and AMN. They become more proactive and started proposing new approaches to the actual issues that the managing team was facing and that CEO did not even imagine himself.	The contextual drivership protocols were explained during the training sessions in order to have all the managers sharing and using easy to use managerial tools. Participants were trained to assess people’s autonomy, to adapt their management style to colleagues’ autonomy and to act consistently, to reward virtuous behaviors and to reinforce positive contributions. Each participant applied the previous tools to produce a mapping of his features and managerial competencies, then he shared this information with every colleague he was sharing an interface with. This allowed all participants to clarify their positioning in the company, their complementary and distinctive competencies, and their management approaches and needs.
With regard to those participants that were more resistant at the beginning, from one side they were “constrained” by the group pressure to use the tools, and on the other side they found in them a tool to explain their and others’ behaviors (positive and negative). This increased their level of comfort and reduced their anxiety of learning. Overall, the managing team was not only learning a common language (the contextual drivership protocols), but as well a meta-language (creating a set of common references to speak about they way they were implementing the tools) (Watzlawick, Beavin, and Jackson,, 1967; Watzlawick, 1983, Tissier, 1988).	During this part of the training, CEO was the one who was challenged the most, and sometimes destabilized, by the learning experience. At the third of five seminars, CEO understood that he had to give up his resistance and turn around his way of managing: he announced it and put it in place not only in the next two sessions, but as well in his day-to-day relationship with his managing team. The last two seminars have been used to fine tune the way the managing team would have implemented the contextual drivership protocols for the next nine months in order to prepare the third part of the training program, the cooperative innovation module.

Table E

The main skills that a learning organization has to develop and enhance to enter a virtuous circle and leverage professional intellect are - Boost professionals' systematic problem-solving abilities by capturing knowledge in systems and software; - Experimentation with new approaches by leveraging tacit knowledge and interdisciplinary cross-functional projects; - Overcome professionals' resistance to share information to facilitate learning from own and others' experience and transfer knowledge and best practices quickly and efficiently throughout the organization (Garvin, 1993; Quinn, Anderson and Finkelstein, 1996).	The seminars on cooperative innovation (June 2003) should have been aimed at: - Encouraging forward thinking, knowledge sharing and best practices seeding; - Creating and supporting the reflex of positive contribution to focus on solutions instead of problems; - Implementing innovative work team technologies and processes; - Producing efficient, effective, simple, feasible solutions to complex issues. Benchmarking internal and external to AMN and EBU as well as best practices sharing meetings will be carried out under the supervision of each manager in order to foster knowledge acquisition capitalization and circulation.

Table F

This time is necessary to stabilize the change, create the reflex and the habit to the use of new tools, and make the tools reliable enough to be spread throughout the organization. The fact that the two bottom managerial levels will start their training around eighteen months after the managing committee can be very useful because: - The lower managers are then trained to tools that they already know "experientially," since their managers already used them in their day-to-day action; - The upper managers are refreshed in their use of the tools by the feed-back of their colleagues. And since they are a step ahead in the process, they should be able to make their colleagues experience new tools.	After the completion of the cooperative innovation module, the managing team should have six months to put in place the tools and fine tuning their use. In December 2003 all the middle and shop floor managers should have passed through the same process and been trained to the same tools the managing team already learned (adapted of course to their level and their needs). During all the implementation period (June 2003-June 2004) the facilitator should have regularly been visiting the company and providing collective and individual coaching to the managing team people in order to help (and sometimes constrain) them to apply in a proper way the tools in their day-to-day action.

4.2.3 FS's Context in March 2003

With regard to the impact of the program in terms of tangible and intangible performance, after one year, the most interesting outcomes observed were:

- Greater cohesiveness of FS's managing team and increased capacity of responding to clients and head-quarter pressure;
- Conflicts between younger and older managers were dramatically reduced and their relationships were much better: young people were rewarded both financially and psychologically, and all but one decided to stay in order to achieve the turnaround change process in their departments; old people were confirmed in their position and contributed to the increase in production and to the cost-cutting program necessary to increase the profitability and the cash-flow;
- Sales and operations staffs met weekly to coordinate actions, proactively anticipating issues and agreeing on a co-build production planning. Including salespeople in the training process helped them to better understand FS's operations and issues, and to take them into account when dealing with aggressive and demanding clients. On the other side, FS's managers improved their understanding of the sales department's functioning and became better at anticipating production issues. AMN and EBU pressure on FS upper management and interfering in their decision making greatly diminished because of FS managing team's greater cohesiveness;
- Despite the downturn in the worldwide economy, overall FS did not suffer much in social and economical terms;

- Concerning employment, FS did not hire interim workers from December 2002 to February 2003, but was able to preserve all its permanent workers;
- FS managed to increase its production and turnover thanks to new contracts, turnaround profitability, positive cash flow, and sharply reduced inventory levels. FS did not wish to disclose its financial and operational figures, which could show what has been the impact of the program on its performance. However, at the beginning of March 2003, CEO stated that in 2002 FS managed to stop negative profitability and provided exceeding cash flow to help AMN to pay back its debt and to sharply reduce the level of inventories;
- CEO convinced EBU to separate personnel administration and training from finance and administration, and to hire a human resource manager for FS.

Still, some issues need to be addressed; as of March 2003:

- The operations of the new six-color printing machine are not optimized, both in terms of production (high levels of scrap) and quality (low precision in printing);
- Low quality in box production due to the increase in volume on old machines (poor preventive maintenance because of lack of investments in previous years);
- Poor control on cost and operations drivers and consequent lack of clarity on priorities to improve production and financial performance.

4.2.4 Using FS's Managers Perception to Analyze the Outcomes of the Implementation of the M3C

It is not possible to say that the positive outcomes observed were all directly related only to the implementation of the model of cooperative contextual change. The analysis of the questionnaire's ratings evolution (Table 4.2.1) showed that FS's managers feel that the training helped them to significantly improve their way of leveraging their competencies and resources. FS's managers coordinate their action, proactively anticipate issues, share information better by speaking a common managerial language, and look for solutions instead of focusing on problems.

The questionnaire asked the participants in the training program to express their perception of the way they were used to working together by rating each question on a scale from zero (low) to nine (high). A color code helped to create clusters according to the request of FS's CEO: any rate over six would have been considered as positive (green); any rate below five would have been considered as highly negative (red); any rate between five and six would have been considered as negative (orange).

The analysis of the evolution of the perception of the participants (Table 4.2.1) provided some interesting insights:

- Twenty-five percent increase (from 4.82 to 6.02) in the mean of the means: A global positive change occurred in the way the managing committee members work together and the way they manage FS and its relationships with AMN and EBU.

- Stability of the standard deviation of the mean of the means (from 0.67 to 0.65): The increase in the average of averages was not due to the dramatic increase of the perception of only some members but to a global, upward, homogeneous shift of the perception, which showed that the managing committee members moved all in the same direction. Such change was homogeneous for all the issues and for the perception that every person had.

		Jun.02		Oct 2002		improvement
		average of data	standard deviation of data	average of data	standard deviation of data	
1	The respect of function's responsibilities in teamwork	5,58	2,23	6,73	1,80	20,7%
2	The distortion of functions' interfaces because of interpersonal conflicts	4,62	2,16	6,15	1,25	33,3%
3	The respect of diversity and differences	5,38	1,65	6,73	1,51	25,0%
4	The capability of making constructive criticism and proposing improvement suggestions	6,15	2,32	6,54	1,56	6,3%
5	the capability of actively listening others' improvement suggestions	5,58	1,74	6,54	1,56	17,2%
6	the capability of recognizing the positive contribution of each individual	5,21	2,38	6,35	1,59	21,8%
7	the collective setting of ground rules and interfaces' rules	3,65	2,32	5,58	2,00	52,6%
8	the respect of ground rules and interfaces' rules	4,04	2,31	5,00	1,39	23,8%
9	the capability of sanctioning when rules are not respected	4,04	2,69	4,81	2,29	19,0%
10	the clarity of the organization and of the strategy of AMN	4,04	1,56	5,00	1,70	23,8%
11	the clarity and the transparency of the relationships between AMN, EBU and FS	4,42	2,00	5,38	1,92	21,7%
12	the clarity of the organization of FS	5,58	2,23	6,35	1,59	13,8%
13	the clarity of my mission within the organization & of my hierarchical, functional and cross-functional relationships	5,38	1,65	7,12	1,92	32,1%
14	the transparency of the strategies of FS and of its implementation action plans	4,42	1,74	6,15	1,59	39,1%
15	the clarity and the sharing of strategic vision	4,23	1,80	5,38	1,92	27,3%
16	the clarity and the sharing of operational actions to be carried out	4,42	1,44	5,58	1,44	26,1%
17	the level of information sharing within the managing team of FS	5,68	1,87	7,00	1,50	23,2%
18	the functioning of the meetings of the managing committee of FS	4,00	2,00	6,50	1,66	62,5%
19	the clarity and the exactness of project processes	5,00	2,64	6,36	1,24	27,3%
20	the soundness and effectiveness of decision-making processes	4,17	2,12	5,77	1,51	38,5%
21	the respect of decisions made	4,81	1,82	6,73	1,15	40,0%
22	the stability and the long-lasting of decisions made with regards to AMN and EBU	4,58	2,00	5,58	1,74	21,7%
23	the quality of decisions' implementation follow-up processes	5,19	2,07	5,58	1,44	7,4%
24	The respect of timing in decisions implementation	5,38	2,16	5,58	1,44	3,6%
	average of averages	**4,82**		**6,02**		**25,0%**
	averages standard deviation	0,67		**0,65**		
	data standard deviation	0,86		**0,70**		

Table 4.2.1—FS's managers perception on managing committee teamwork effectiveness

- Increase in the mean of each item: It shows an improvement in the perception that the members of the managing committee have of the way they work together in all the fields that (during the preliminary audit) they considered as critical for their performance. As shown by the color code, in June 2002 thirteen items over twenty-four were red, ten were orange, and only item four was green (constructive criticism, that means that in a certain way they were ready for a change process aimed to question and turn around the way they produce and use knowledge) globally; a few months later, in October 2002, thirteen items were green, ten were orange and only one was red (item nine, the capability of sanctioning when rules are not respected, that means that they still have to work on group discipline). No item showed a decrease. Looking at the detail, in four items they perceived a two-step improvement (from red to green in items 2, 14, 18, 21), in fourteen items they perceive a one-step improvement (from red to orange in items 7, 8, 10, 11, 15, 16, 20, 22; from orange to green in items 5, 6, 12, 13, 17, 19), and only in three items they do not perceive a step improvement (stable orange for items 23, 24; stable red for item 9);
- Decrease or stability (for all items but 13, 15, and 16) of the standard deviation of the mean of each item: Not only the perception of people improved, but people are closer to each other in their perception as well;
- Decrease of the data standard deviation (from 0.86 to 0.70): The cohesiveness of the team globally increased.

In November and December 2002, several members of the managing team accepted to pass non-directive interviews to express their qualitative perception of the short-medium term effects of the implementation of the first two modules of the model of cooperative contextual change. The table below provides some of their most interesting comments (right-hand column) and the author's observations or analysis (left-hand column) concerning the main changes people observed in teamwork-related issues and management tools implementation and results.

Table G

The model of cooperative contextual change helped the managers to create the most favorable conditions to work more effectively together. Change and stability are then the consequence of its implementation.	"The improvement of our performance did not happen because of the methods and tools we learnt during the training: they are just good sense, things that a good manager usually does intuitively. But the program made us become conscious of the context, it helped us to better know each other, to have a good team dynamics, and to increase our team cohesiveness" (Fred, systems and maintenance manager). "We are more dispassionate in our day-to-day operations relationship; before we were wasting so many energies in conflicts and disputes. There are no longer polemics…but a greater capability of working constructively together. Now we are a team" (Monique, sales manager).

Table H

The model of cooperative contextual change has been designed to integrate three main approaches (Blake and Mouton, 1985, p.219-222): - Techniques-based approach; that is, providing managers with new managerial tools and techniques, in order to provoke change 1 (Watzlawick, Weakland and Fisch, 1974); - Operating procedures-based approach; that is, leveraging the contribution of an external trainer to make explicit and modify individual and team way of behaving in order to provoke change two (Watzlawick, Weakland and Fisch, 1974); - Theoretical knowledge-based approach; that is, providing managers with pragmatic models that can explain the outcomes of their behaviors and be simple enough to be used in their day-to-day operations to adopt a proactive attitude toward change two.	"At the beginning we thought that the training program would have provided us with nothing but tools that we were already using in our day-to-day professional life. In fact, we learned how to work differently to make reference to a method" (Monique, sales manager). "Most of the tools are easy to use: commitment, competencies, reliability, autonomy…and it is not necessary to write them: what matters is to have them in mind" (Etienne, products development manager). "Many of us have changed and are experiencing a real evolution. Today is easier to have positive constructive relationships with most of my colleagues" (Etienne, products design manager). "Some of us understood the importance and the impact of the change process and want to go ahead to improve our knowledge of the method and the understanding of our way of doing" (Annie, quality manager).

Table I

Even if the implementation of the model of cooperative contextual change had been specially designed to fit FS's context, it could not meet everybody's needs and expectations; as any change process, it provoked resistance because it aimed to shake old habits and modify long-lasting way of behaving. What matters is to accept resistance as an opportunity to understand new features of the context and in this way find out the way to overcome it (Haley, 1973, 1976).	"There are some people who are not ready for change yet who did not understand and accept the new way of working. Maybe, individual coaching would be beneficial to them" (Etienne, products design manager). "There are still a few people who did not understand, or do not want to. Other struggle to clarify what is at stake for them in this change" (Annie, quality manager).

Table J

The implementation of the model of cooperative contextual change provoked: - An increase in the level of comfort, especially in what concerns belongingness (team social dynamics), but as well in safety (more positive criticism, reciprocal help when facing difficulties) and physiological needs (better working environment and positive use of energies); - The enactment of stimulation factors as self and reciprocal esteem and recognition of individuals contribution, as well as opportunities for personal and professional development. The simultaneous improvement of the comfort and the stimulation dimensions provoked an increase in the level of commitment of the members of the managing committee, as witnessed by the dramatic change in the questionnaire, which measures the perception they had of their way of behaving and working together. The consequences are: - A positive change in their operating processes (like meetings, information sharing, shop floor production); - A reduction in conflicts and a better use of their resources to handle operational issues (creative problem solving and solution seeking); - An increase in the cohesiveness of the team and in the capability of using a common language, that is to share a common vision (Senge et al., 1994).	"Commitment increased: today everybody pushes in the same direction, we behave differently from the past, we are more positive" (Monique, sales manager). "We all use the same words and have the same reflexes of respectful contribution. We support and help each other. We understood that disputes are useless" (Henri, logistics manager). "Today we are able to perceive issues correctly, to make them clear, to formalize them. Now we have to look for solutions" (Etienne, products development manager). "Everybody accepts criticism and questions himself. Now we have to change our organization to take advantage of our new way of working" (Etienne, products development manager). "Our meetings, especially the weekly meeting of the managing team, are faster, and we share higher quality information in a more effective way" (Marius, production manager). "Before, we were used to focusing on problems, and we had conflicts between departments and with the younger managers. Today everybody is more explicit and we look for solutions" (Louis, printing workshop manager).

Table K

The effectiveness of a cooperative contextual change program depends a great deal on the setting and enforcement of ground rules. Therefore, one of the key factors of success is the use of sanctions to make sure that the ground rules are correctly implemented. It is especially important that the driver follows and enforces the rules, for himself and for his colleagues, since it constitutes an example and a major sign of commitment.	"Applying correctly positive and negative sanctions is paramount. We as managing committee we should provide the example; we can demand more to our CEO: he attended the same training, he must be the first" (Paul, financial and administrative manager). "The CEO should be more explicit on what is at stake and formalize more what are purposes and goals for each function; we need more concrete actions and to gear the program down into the organization" (Enzo, production coordinator).

Table L

A coherent implementation of the contextual drivership protocols allows people to increase the efficiency and effectiveness of their actions, focus more on what is important when it is not urgent in order to anticipate issues, and be proactive when critical events happen. All these things contribute to the increase of the stability of the individuals and of the teams, and therefore enhance the performance of the organization and the long-lasting of change.	"Thanks to the manager's added value matrix, we can now work on the substance of issues, better shape our core activities, and look for scenarios to anticipate solutions" (Etienne, products development manager). "Today we are able to better manage crises without being stuck in the middle" (Paul, financial and administrative manager). "With regard to the shop floor production activities, crises are managed much better than before" (Annie, quality manager).

Table M

As for any systemic approach to change, the implementation of a cooperative contextual change program affects all the actors of a company, since it produces changes in the behaviors of interdependent elements. Moreover, as we have seen above, it introduces some invariants in the functioning of the organization (e.g., ground rules, rituals, processes, terminology); these invariants deeply affect the relationships between people, because they become a meta-language which allows effective meta-communication (Watzlawick, Beavin, Jackson, 1967; Watzlawick, Weakland and Fisch, 1974; Tissier, 1988). It is normal, therefore, that the managers ask for the program to be geared down into their organizations, in order to impulse and speed up the change at any level of the company. At the same time, it is important that they recognize the usefulness of an external person as the trainer who can help to catalyze the process of change by facilitating the process of making the context explicit and acting as a mediator to orient energies toward the solution (Malarewicz, 2000).	"It is necessary to gear the program down to the floor shop line managers" (Monique, sales manager). "The program should be geared down up to the operators. Some tools (reliability matrix, autonomy curve) could be explained by us to our people. Others should be delivered by the external trainer, as the community norming and interpersonal communication tools" (Annie, quality manager). "For the program to be effective and long lasting, it should be geared down to the shop floor operators. And it should be done by an external trainer in order to avoid hierarchy issues and unnecessary resistance" (Enzo, production coordinator).

4.2.5 Unexpected Findings

Until the beginning of March 2003, information was received that the M3C and the drivership protocols were still being implemented at FS. As reported above, FS's CEO and members of the managing team saw significant improvement in the performance of the subsidiary, which they attributed to the implementation of M3C.

However, at the end of April 2003, the CEO asked for an investigation of the difficulties of the subsidiary. When visiting the company, in discussion with the CEO, he discovered that in the last two months, the CEO's attention had been diverted to the annual salary negotiation. This diversion of attention reduced the emphasis on the continued implementation of M3C and the drivership protocols. The CEO stopped being the example for the implementation of the change process and did not continue to monitor the activities that would protect the stake. As a consequence, the management committee members did not continue to enforce the ground rules and to apply the drivership protocols. This led to the worsening of the relationship among them. As the interpersonal relationships deteriorated, conflicts among departments increased, and the performance of the subsidiary worsened.

The salary package offered as part of the negotiation was not comparable to what was offered by other subsidiaries due to FS's sudden downturn in performance. The trade unions called then for an unexpected strike, protesting the deterioration of the working relationships within the subsidiary. The trade unions did not put strikes in place in other AMN subsidiaries. The CEO decided to be very confrontational with the trade unions, assigning the workers to the court and exacerbating internal conflicts. The strikes stopped, but this tense situation lasted until June 2003, with further deterioration of FS's performance. In September 2003, the CEO was fired.

CONCLUSIONS

The purpose of the model of cooperative contextual change (M3C) is to produce sustainable, effective improvement in a system. The model suggests that individuals involved in change need to experience dynamic stability. Individuals need to have a sense of community and responsibility for protecting the explicit common stake through ground rules. Individuals must commit to accepting responsibility for the initiating and regulating actions for protecting the stake and must look for solutions to improve the current situation.

The two case studies provided the opportunity to test these theses.

Conclusion One

The RC case supported the belief that the autonomy curve is an effective tool for assessing the capacity of a driver to accept responsibility and engage in actions that protect the stake.

Based on what was observed in the training, we believed that Ms. Papin exhibited the characteristics of an autonomous individual. After the training, Ms. Papin was given by RC's CEO the position of industrial director, with the responsibility of turning around the operational performance of the company. Ms. Papin understood and internalized the fact that her actions were a necessary contribution to protect the stake of the company to save RC's French operations.

At the end of eighteen months, the performance of the company had improved significantly, primarily because of the improvement in the operational and financial performance of the industrial department. In regard to the improvement of internal processes,

Ms. Papin cited increased coordination of people's actions, better capability to proactively anticipate issues, enhanced sharing of information by speaking a common managerial language, willingness to look for solutions instead of focusing on problems, and higher readiness to face future changes in a more cooperative way.

Conclusion Two

As RC's industrial director, Ms. Papin accepted the need for installing the four elements of dynamic stability as a condition to implement a successful change process.

After analyzing the conditions within the company, she recognized that to ask for significant immediate changes would result in an increase in the stress and instability. Therefore, she staged her actions as follows:

- First, she selected one improvement goal, the reduction to zero of the late items, which established a stake acceptable by the employees of the industrial activities department;
- Second, she asked the employees to co-build and co-opt ground rules in order to achieve the reduction of the late items;
- Third, she customized the responsibilities given to those who would enact it to allow them to appropriately initiate and regulate actions;
- Fourth, she asked the employees to identify solutions to accomplish the goal within a time frame they suggested and that she agreed upon.

The employees reached this goal prior to the end of the time frame. This allowed Ms. Papin to move to other more ambitious and complex goals, because the employees were involved in the process and were more willing to be involved in finding solution to further enhance the change process.

The successful implementation of the M3C at RC demonstrates the criticalness for the change driver to leverage commitment, passion, patience, intelligence, and legitimacy.

Conclusion Three

The FS case provides the evidence that failing to internalize the principles of M3C and drivership, as well as a lack of consistency and perseverance in their application may jeopardize the effectiveness and successfulness of the change process.

The shifting of the CEO attention to what he believed were more important issues subverted the internalization of the M3C. Thus, the four elements of stability and the drivership protocols were not established as the mindset of reference for the actors of the change process. When difficulties arose, all the actors (the CEO, the managers, and the employees) reverted to their old mental models of relationships, demonstrating that they had not reached the appropriate level of autonomy necessary to drive and sustain the change process.

Conclusion Four

The appropriate implementation of the M3C and the drivership protocols cannot be seen as the only reason for the improvement of RC's performance, nor their inappropriate implementation causing the worsening of FS's performance. However, the cases show that the M3C and the drivership protocol can have a substantial impact on creating a sense of community among the actors of the change. The sense of community establishes the conditions, which can enhance cooperation.

The most significant barrier to a successful change initiative is a lack of cohesiveness between people and of consistency in their individual and collective decision-making and actions. Fostering cooperation may be priceless in designing and managing a strategic and/or organizational turnaround (Hersey and Blanchard, 1972). Cooperation constitutes a major factor of stability, especially when people have to protect several (and sometimes divergent) high stakes in a stressing context, as it is common in contemporary global business environment. Encouraging the emergence of cooperation is one of the most critical issues leaders and managers face when confronted to change issues (Axelrod, 1984, Nalebuff and Brandenburger, 1996).

RECOMMENDATIONS

The recommendations that follow are based upon the assessment of three primary experiences:

- Involvement in change processes that both succeeded and failed;
- Training people on the components of the M3C as well as the model itself, i.e., the RC case;
- Providing consultation to support change initiatives based on the M3C, e.g., the FS case.

Recommendation One

The change initiator must use a model that fits the complexity of the context. This means that the model must provide the tools for an appropriate analysis of the interdependence of the elements of the system, which is the source of the complexity. At the same time, the model must suggest an implementation process that can be easily understood by the change actors, who are the source of effectiveness.

When an initiator of change is faced with a complex issue whose outcome is uncertain, under the perception that time is of the essence, the tendency is to seek a quick answer using a simple model. Therefore, insufficient time is taken to analyze all the different factors and elements within the context. This simplistic analysis logically leads to a simplistic answer, which is often not effective. Now the change initiator must move even more quickly to find another solution and select the next easiest answer.

The change initiator must recognize that time spent in carefully obtaining different perspectives within the organization from people at different levels will actually shorten the time to resolving the issue. Involving the individuals, who will enact the change process, in describing the context and offering solution creates awareness and ownership, which are the foundations of success.

Recommendation Two

If conflicts exist among the individuals who will be asked to enact change, these emotional conflicts must be resolved before moving to the next step of the change process.

As the context's analysis is carried out, the change initiator may discover that interpersonal conflicts exist among the change actors. If these interpersonal conflicts are not resolved before moving to the implementation of the change process, the entire process will be jeopardized. People experiencing emotional conflicts will divert their energies to the conflict away from the change initiative.

Recommendation Three

The change initiator must alter his mindset regarding the need to induce instability as a requirement for change. Instead, he must understand the need of stabilizing the perceptions of the individuals with regard to the dynamics of change.

When the need for an intervention is recognized, change initiators tend to accept as a fact that instability is going to occur because individuals have to give up the status quo. Unfortunately, if instability is introduced within a stable system without providing emotional and intellectual anchors, the consequences of instability itself become the reason for failure.

The perception of stability occurs when the change actors are involved in co-building and co-opting the ground rules that will support the activities to protect the stake. In this manner, the change initiator has provided a perception of ownership to the change actors.

Recommendation Four

The change initiator must assess the level of commitment and the level of competence of those who will be responsible for enacting the change initiative, in regard to what is defined as "level of autonomy."

The change initiator must resist the temptation of treating everybody in the same way, because it seems fair, which assumes all have the same level of competence and commitment. He must understand the difference between giving respect to the person and assigning responsibilities based upon the level of autonomy for a given task in a given context.

Recommendation Five

The change initiator must continue to support, encourage, and monitor the implementation of the M3C and drivership protocols until there is evidence that the four elements of dynamic stability have been internalized by the change actors.

When change initiators begin to experience success with a change initiative, there is a tendency to focus their attention on what appear to be more pressing issues. If the change actors have not internalized the M3C and the drivership protocols, they may begin to deviate from the agreed behaviors. These negative consequences can lead to the subversion and downfall of the change initiative. Therefore, the change initiator must be sure that there is a sufficient level of internalization, which is exhibited through the change actors' level of autonomy.

Recommendation Six

The M3C and the drivership protocols should be tested in a variety of industrial settings and cultural contexts to prove its validity.

APPENDICES

Type	Degree of pain felt	Primary motivation	Degree of threat to survival	Gap between environmental needs and operations	Clarity of outcome	Impact on mindset	Focus of change	Orientation	Level of personal development required	How change occurs
Developmental Change	1	Improvement	1	1	4 It is prescribed against a standard	1 Little if any	Improvement of skills, knowledge, practice, and performance	To do better in a certain area: project oriented	1	Through training, skills development, communication, process improvement
Transitional Change	2	Fix a problem	2	2	4 It is designed against a criteria	1 Little if any	Redesign of strategy, structures, systems, processes, technology or work practices (not culture)	Project-oriented; largely focused on structure, technology, and work practices	2	Controlled process, support structures, timeline
Transformational Change	3-4	Survival: change or die or breakthrough needed to pursue new opportunities	1-4	3-4	1 It is not initially known; it emerges or is created through trial and error and continuous course correction	2-4 Forced to shift, old mindset and/or business paradigm mindset must change	Overhaul of strategy, structure, systems processes, technology, work, culture, behavior and mindset	Process-oriented requires shift in mindset, behaviors, and culture	3-4	Conscious design and facilitation; high involvement; emergent process

Appendix One—Features of the Different Types of Change

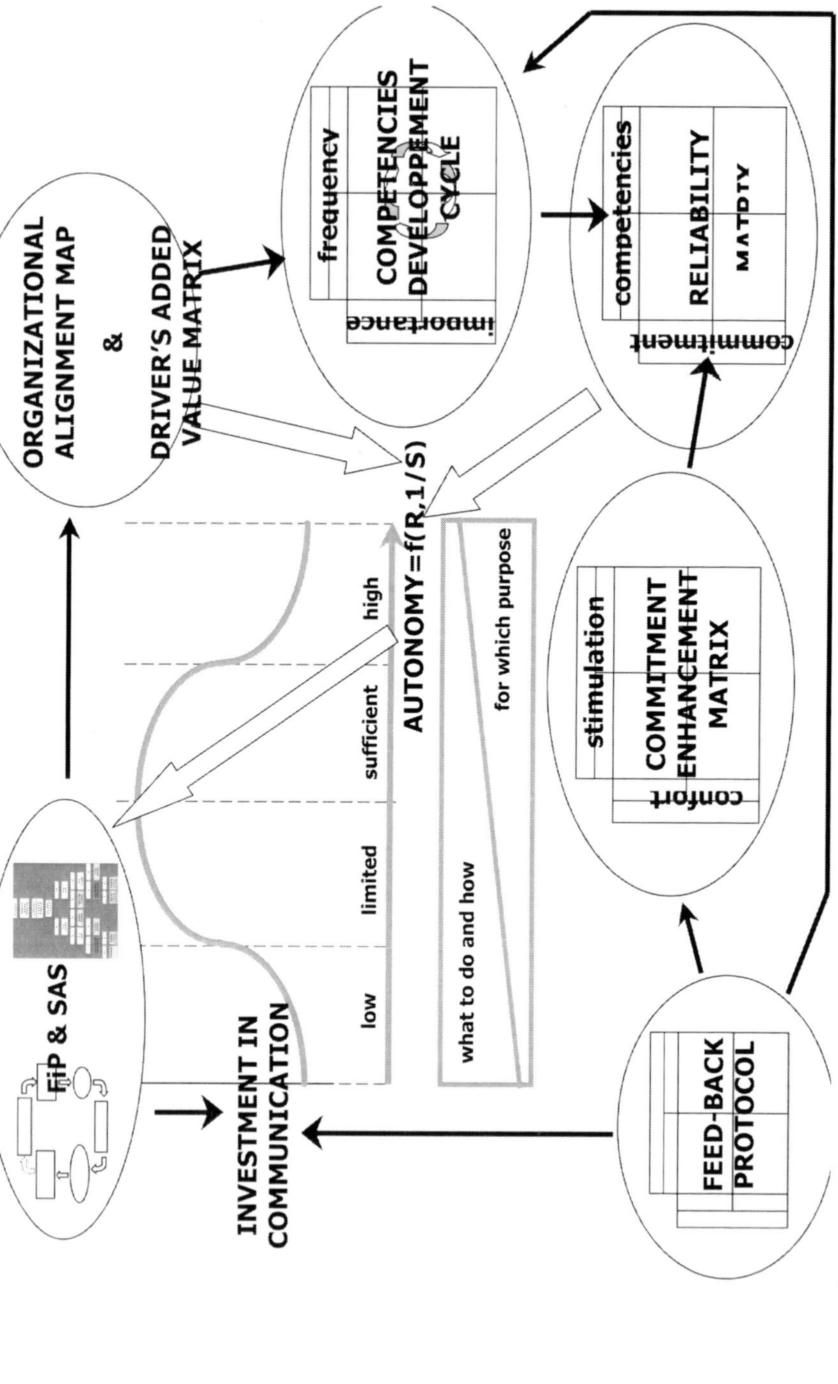

Appendix Two—Global View of the drivership Approach

BIBLIOGRAPHY

Adler, cited by Choo, C.W. and Bontis, N. (2002a) Knowledge, Intellectual Capital, and Strategy: Themes and Tensions. In Choo, C.W. and Bontis, N. (Eds.), The Strategic Management of Intellectual Capital and Organizational Knowledge. Oxford University Press, New York

Allen, N.J. and Meyer, J.P (1990) The Measurement and Antecedents of Effective, Continuance and Normative Commitment to the Organization. Journal of Occupational Psychology, 63, 1–18

Alter, N. (1993) Innovation et Organization, Deux Légitimités en Concurrence. Revue Française de Sociologie, XXXIX, 2, 175–197

Anderson, D.; Ackerman Anderson, L. (2001) Beyond Change Management. Jossey-Bass, Pfeiffer, San Francisco

Arendt, C.H.; Landis, R.M. and Meister, T.B. (1995) The Human Side of Change, Part Four. IEE Solutions, 22–6

Argyris, C. (1991) Teaching Smart People How to Learn. Harvard Business Review, May-June

Argyris, C. (1995) Savoir pour Agir : Surmonter les Obstacles à L'Apprentissage Organisationnel. InterEditions, Paris

Aubret, J. (1993) Savoir et Pouvoir, les Competences en Questions. PUF, Paris

Aubrey, B. (1994) Le Travail Après la Crise Paris. InterEditions, Paris

Aubrey, B. (1999) What is Management? Handouts of the Professional Development Seminar for Professors' Training. Reims Management School, March

Axelrod, R. (1984) The Evolution of Cooperation. Basic Books

Barjou, B. (1999) Manager par Projet. Methodes et Comportements pour Animer Hors Statut Hierarchique. 2ème edition. ESF Editeur, Paris

Bass, B.M. (1960) Leadership, Psychology, and Organization Behavior. Harper and Bros, New York

Bass, B.M. (1985) Leadership and Performance Beyond Expectations. Free Press, New York

Baum, G. (1977) The Logic and Limits of Trust. Rutgers University Press, New Brunswick, NJ

Baumard, P. (1996) Organisations Deconcertées: la Gestion Stratégique de la Connaissance. Masson, Paris

Becker, H.S. (1960) Notes on the concept of commitment. American Journal of Sociology, 66, 32–40

Beckhard, R and Harris, R. (1987) Agent of Change: My Life, My Practice. Jossey-Bass, San Francisco

Beer, M. and Nohria, N. (2000) Cracking the Code of Change. Harvard Business Review, May-June, 132–141

Bennis, W.G. (1989) Managing the Dream: Leadership in the 21st Century. Journal of Organizational Change Management, V01.2 Issue 1, 6

Benson, J.K. (1977) Organizations: a Dialectical View. Administrative Science Quarterly, Vol. 22, 1–19

Berson, Y.; Shamir, B.; Avolio, B.J. and Popper, M. (2002) The Relationship between Vision Strength, Leadership Style, and Context. The Leadership Quarterly, 12, 53–73

Biscaccianti, A. and Feron, M. (1999) Focusing on Capabilities Development to Reduce Knowledge and Skills Obsolescence: A French Experience in the Management School System.

In Theile, K. and O. Hogartaigh, C. (editors), International Business Education. Partnership, Patterns and Prospects for the 21st Century. Oak Tree Press, Dublin

Blake, R.R. and Mouton, J.S. (1985) The Managerial Grid 3rd: The Key to Leadership Excellence. Gulf Publishing Company, Houston

Blanchard, K. and Johnson, S. (1981) The One Minute Manager. 1981

Bovey, W.H. and Hede, A. (2001a) Resistance to Organizational Change: The Role of Cognitive and Affective Processes. Leadership and Organization Development Journal, 22(8)

Bovey, W.H. and Hede, A. (2001b) Resistance to Organizational Change: The Role of Defense Mechanisms. Journal of Management Psychology, 16, 7

Brouwers, I.; Cornet, A.; Gutierrez, L.; Pichault, F.; Rousseau, A. and Warnotte, G. (1997) Management Humain et Contexte de Changement. Pour une Approche Constructiviste. De Boeck Université, Paris-Bruxelles

Brown, M.G. and Eisenhardt, K.M. (1995) Product development: Past Research, Present Findings, and Future Directions. Academy of Management Review, 20, 343–378

Buchanan, B. (1974) Building Organizational Commitment: The Socialization of Managers in Work Organizations

Carr, C. (1994) Seven Keys to Successful Change, Training, 31, 2, 55–60

Chandler, A. (1973) Decision Making and Modern Institutional Change. Journal of Economic History, Mar, n. 33, 1–15

Chandler, A. (1976) The Development of Modern Management Structure in the US and UK. In L. Hannah (ed.), Management

Strategy and Business Development: An Historical and Comparative Study. McMillan, London, 23–51

Chandler, A. (1989) Stratégies et Structures de L'Entreprise. Les Editions D'Organisation, Paris

Choo, C.W. and Bontis, N. (2002a) Knowledge, Intellectual Capital, and Strategy: Themes and Tensions. In Choo, C.W. and Bontis, N. (Eds.), The Strategic Management of Intellectual Capital and Organizational Knowledge. Oxford University Press, New York

Choo, C.W. and Bontis, N. (Eds.) (2002b) The Strategic Management of Intellectual Capital and Organizational Knowledge. Oxford University Press, New York, 3–19

Cialdini, R.B. (2001) Influence. Science and Practice. Allyn and Bacon. Needham Heights

Coch, L. and French, J.R.P. (1948) Overcoming Resistance to Change. Human Relations, 1, 512–532

Cohen, W. M., and Levinthal, D. A. (1990). Absorptive Capacity: A New Perspective on Learning and Innovation. Administrative Science Quarterly, 35, 128–152.

Collins, J. (2001) Good to Great. Fast Company, Oct, issue 51, p90–100

Conger, J.A. and Kanungo, R.N.(1998) Charismatic Leadership in Organizations. Sage, Thousands Oaks, CA

Crozier, M. and Friedberg, E. (1977) L'Acteur et le Système. Edition Point Seuil, Paris

Crozier, M. (1963) Le Phénoméne Bureaucratique. Seuil, Paris

Csikszentmihalyi, M. (1991) Implication of a Systems Perspective for the Study of Creativity. In Sternberg, R.J. (ed.), Handbook of Creativity. Cambridge University Press, Cambridge, 313–335

De Board, R. (1978) The Psychoanalysis of Organizations. Tavistock, London

De Board, R. (1983) Counseling Skills. Gower Publishing

De Bono, E. (1985) Six Thinking Hats. Key Porter Books Ltd, Toronto

De Dreu, C.K.W. and Beersma, B. (2001) Minority influence in organizations: Its Origins and Implication for Learning and Group Performance. In de Dreu, C.K.W. and de Vries, M.K. (eds), Group Consensus and Minority Influence. Blackwell, Oxford, pp.258,283

De Saint Paul, J. and Tenenbaum, S. (2002) L'Esprit de la Magie. La Programmation Neuro-Linguistique. Relation à Soi, Relation à L'Autre, Relation au Monde. InterEditions-Dunod, Paris

Deci, E. L. (1992) The History of Motivation. In Vroom,V.H. and Deci, E.L (Eds.) Psychology and Its Relevance for Management and Motivation. Penguin, London

Demers, C. (1999) De la Gestion du Changement à la Capacité de changer: L'Evolution de la Recherché sur le Changement Organisationnel de 1945 à Aujourd'hui. Gestion, vol. 24 n.3

Denison, D.R. and Mishra, A.K. (1995) Toward a Theory of Organizational Culture and Effectiveness. Organization Science: A Journal of the Institute of Management Sciences; Mar-Apr, vol. 6 issue 2, 204–224

Desikachar, K. (1999) Stability in Change. Notes of the Seminar. Association Viniyoga, Reims

Drucker, P. (1954) The Practice of Management.

Durkheim, E. (1893, 1984) The Division of Labor in Society (Halls, W.D., trans.). Free Press, New York

Erickson, M.H. (1967) Advanced Techniques of Hypnosis and Therapy: Selected Papers of Milton H. Erickson. Edited by Haley, J. Grune and Srtatton, New York

Erickson, M.H. (1980) The Collected Papers of Milton H. Erickson on Hypnosis. Edited by Rossi, E.L.; The Nature of Hypnosis and Suggestion, Vol 1. Irvington, New York

Erickson, M.H. (1992) Healing in Hypnosis. The Seminars, Workshops and Lectures of Milton H. Erickson. Edited by Rossi, E.L.; Ryan, M.O. and Sharp, F.A. Free Associatio Books, London

Etzioni, A. (1961) A comparative Analysis of Complex Organizations. The Free Press of Glencoe, New York

Ezzamel, M. and Willmott, H. (1998) Accounting for Teamwork: A Critical Study of Group-based Systems of Organizational Control. Administrative Science Quarterly, 43, 499–514

Fisher, R., Ury, W. and Patton, B. (1981, 1991) Getting to Yes. Houghton Mifflin Company, Boston

Ford, C.M. (2000) Creative Developments in Creativity Theory. Academy of Management Review, 25, 284–285

Freeman, R.E. (1984) Strategic Management: A Stakeholder Approach. Pitman, Boston

French, J.R.P and Raven, B. (1959) The Bases of Social Power. In Cartwright, D. (Ed.) Studies in Social Power. Institute for Social Research, Ann Arbor, MI

Gabriel, Y. (1992) Putting the Organization on the Analyst's Couch. European Management Journal, 10, 3, 348–351

Garvin, D.A. (1993) Building a Learning Organization. Harvard Business Review, Jul-Aug, vol. 96 issue 1, p.19–29

Giddens, A. (1979) Central Problems in Social Theory: Action, Structure, and Contradiction in Social Analysis. Mc Millan, London

Grant, R. M. (1996b) Prospering in Dynamically-competitive Environments: Organizational Capability as Knowledge Integration. Organization Science, 7(4), 375.

Grant, R. M. (1996a) Prospering in Dynamically-competitive Environments: Organizational Capability as Knowledge Integration. Organization Science: A Journal of the Institute of Management Sciences (INFORMS: Institute for Operations Research, Vol. 7, 375

Grant, R. M. (1996c) Toward a Knowledge-based Theory of the Firm. Strategic Management Journal; 17 (Special Winter), 109–122.

Grant, R.M. (2001) Knowledge and Organization. In Nonaka, I. and Teece, D.J. (Eds.), Managing Industrial Knowledge. Sage Publications, London, pp.145–169

Gray, R. ; Dcy, C. ; Owen, D.; Evans, R. and Zadek, S (1997) Struggling with the Praxis of Social Accounting; Accounting, Auditing and Accountability Journal, 10:3, 325–364

Greenleaf, R.K. (1977) Servant Leadership: A Journey into the Nature of Legitimate Power and Greatness. Paulist Press, Mahwah, NJ

Greiner, L. (1998) Evolution and Revolution As Organizations Grow. Harvard Business Review, May-Jun, 55–67

Grimand, A. (1995) L'Emergence de la Notion de Compétence en Gestion des Ressources Humaines : Discours, Pratiques, Enjeux. Actes du 6ème congrès de l'AGRH, pp. 241–244

Hackman, J.R. (1990) Groups That Work (And Those That Don't): Creating Conditions for Effective Teamwork. Jossey-Bass, San Francisco

Haley, J. (1973) Uncommon Therapy. New York, Norton

Haley, J. (1976) Problem-Solving Therapy, New Strategies for Effective Family Therapy. Jossey-Bass, San Francisco

Hall, a.d. and Fagen, R.E. (1956) Definition of System. General Systems Yearbook, vol. 1, 18–28

Handy, C. (1989) The Age of Unreason. Business Books Ltd, London

Harrison, D. and Laberge, M. (2002) Innovation, Identities, and Resistance: The Social Construction of an Innovation Network. Journal of Management Studies, june, 39, 4, 497–522

Hauschildt, J. and Kirchmann, E. (2001) Teamwork for Innovation—The "Troika" of Promotors. RandD Management, 31,1, 41–49

Hebbar, K. (2003) Pratiques Innovantes en Matière de Mobilisation Stratégique des Ressources Humaines: Impact sur la Création de Valeur, la Productivité and la Croissance dans le Cadre de Jeunes Entreprises Technologiques. Colloque SupdeCo Amiens, Octobre

Hersey, P. (1984) The Situational Leader. Center for Leadership Studies, Escondido, CA

Hersey, P. and Blanchard, K H. (1969) Life Cycle Theorie of Leadership. Training and Development Journal, May, 26–34

—-. (1972) The Management of Change, Part1. Training and Development Journal, Jan, 6–10

—-. (1972) The Management of Change, Part2. Training and Development Journal, Feb, 20–24

—-. (1972) The Management of Change, Part3. Training and Development Journal, Mar, 28–33

—-. (1974) What-s Missing in MBO? Management Review, Oct, 25–32

—-. (1996) Revisiting the Life-Cycle Theory of Leadership. Training and Development Journal, Jan, p.42–47

Herzberg, F. (1959) Motivation to Work. John Wiley and Son, New York

Herzberg, F. (1966) Work and the Nature of Man. World Publishing Co, New York

Hill, C.W. (1990) Cooperation, Opportunism, and the Invisible Hand: Implications for Transactions Cost Theory. Academy of Management Review, 15, 500–513

Hosmer, L.T. (1995) Trust: the Connecting Link Between Organizational Theory and Philosophical Ethics. Academy of Management Review, 20:2, 379–403

Hultman, K.E. (1995) Scaling the Wall of Resistance. Training and Development, 49, 10, 15–18

Huse, M. and Eide, D. (1996) Stakeholder Management and the Avoidance of Corporate Control. Business and Society, 32:2, 211–243

Huston, L.A. (1992) Using Total Quality to Put Strategic Intent to Motion. Planning Review, 20, 5, 21–23

Isaken, S.G. and Lauer, K.J. (2002) The Climate for Creativity and Change in Teams. Creativity and Innovation Management, Mar, vol 11, n. 1

Jackson, S.E. (1996) The Consequence of Diversity in Multidisciplinary Work Teams. In West, M.A. (ed.), Handbook of Work Group Psychology. Wiley, Chichester, 53–75

Jacques, E. (1972) Intervention et Changement dans L'Entreprise. Dunod

Jassawalla, A.R. and Sashittal, H.C. (1998) An Examination of Collaboration in High-technology New Product Development Processes. Journal of Product Innovation Management, 15, 237–254

Jassawalla, A.R., Sashittal, H.C. and Hemant, C. (1998) Cultures That Support Product-innovation Processes. Academy of Management Executive, Aug, 16, 3, 42–55

Jehn, K,A., Northcraft, G.B. and Neale, M.A. (1999) Why Differences Make a Difference: A Field Study of Diversity, Conflict, and Performance in Work-groups. Administrative Science Quarterly, 44, 741–763

Johnson, D.W., Maruyama,., Johnson, R.T., Nelson, D. and Skon, L. (1981) Effects of Cooperative, Competitive, and Individualistic Goal Structures on Achievement: A Meta-analysis. Psychological Bulletin, 89, 47–62

Johnson, W.R. and Jones-Johnson, G. (1992) Subjective Underemployement and Psychosocial Stress: The Role of Perceived Social and Supervisor Support. Journal of Social Psychology, Feb, vol 132 issue 1, p11–22

JohnsonD.W. and Johnson, F.P. (2000) Joining Together: Group Theory and Group Skills (Seventh ed). Allyn and Bacon, Boston

Joule, R.-V. and Beauvois, J.-L. (1987) Petit Traité de Manipulation à l'Usage des Honnetes Gens. PUG, Paris

Joule, R.-V. and Beauvois, J.-L. (1998) La Soumission Librement Consentie. PUF, Paris

Kanter, R.M. (1968) Commitment and Social Organization: A Study of Commitment Mechanisms in Utopian Communities. American Sociological Review m 33m, 499–517

Kaplan, R. and Norton, D. (1992) The Balanced Scorecard—Measures that Drive Performance. Harvard Business Review, Jan-Feb, vol. 70 issue 1, 71–80

—-. (1996) Using the Balanced Scorecard as a Strategic Management System. Harvard Business Review; Jan-Feb, p75–85

Kilmann, R.H. and Saxton, M.J. (1983) The Kilmann-SaxtonCulture-Gap Survey. Organizational Design Consultants, Pittsburgh, PA

Klarsfeld, A. and de Saint-Giniez, V. (1999) Construire l'Apprentissage. Ouvertures, Groupe ESSEC, no. 1, Hiver.

Kleiner, A. and Roth, G. (1997) How to Make Experience Your Company's Best Teacher. Harvard Business Review, Sep-Oct

Knights, D. and Wilmott, H.C. (1997) The Hype and Hope of Interdisciplinary Management Studies. British Journal of Management, 8, 9–22

Koestler, A. (1964) The Act of Creation. The Macmillan Company. New York

Kogut, B. and Zander, U. (1992) Knowledge of the Firm, Combinative Capabilities, and the Replication of Technology. Organization Science, 3: 384

Korzybski, A. and Kendig, M. (1942) "Foreword" to General Semantics Monograph III, a Theory of Meaning Analyzed. The International Non-Aristotelian Library Publishing Company, Lakeville, Connecticut

Kotter, J.P (1995) Leading Change: Why Transformation Efforts Fail. Harvard Business Review, Mar-Apr, 59–67

Kotter, J.P (1996) Leading Change. Harvard Business School Press, Boston, MA

Lax, D.A. and Sebenius, J.K. (1986) The Manager as Negotiator. The Free Press

Le Boterf, G. (1994) De la Competence, un Essai sur un Attracteur Etrange. Editions d'Organisation, Paris

Lee, M. and Koh, J. (2001) Is Empowerment Really a New Concept? International Journal of Human Resource Management, 12: Jun, 684–695

Levine, G. (1997) Forging Successful Resistance. Bobbin, 39, 1, 164–166

Lewin, K. (1950) Resolving Social Conflict. Harper, New York

—-.(1951) Field Theory in Social Science. Harper, New York

Likert, R. (1961) New Patterns of Management. McGraw-Hill, New York

—-. (1967) The Human Organization: its Management and Value. McGraw-Hill, New York

Lipton, M. (1996) When clients resist change. Journal of Management Consulting, 9,2,16–21

Liyanage, S. and Barnard, R. (2002). What is the Value of Firms' Prior Knowledge? Building Organizational Knowledge Capabilities. Singapore Management Review, 24(3), 35.

Malarewicz, J. (2000) Systemique et Enterprise. Dunod, Paris

March, J. (1991) Exploration and Exploitation in Organizational Learning. Organization Science, 2: 71–87

Maslow, A. H. (1943) A Theory of Human Motivation. Psychological Review, 50: 370–396

—-. (1968) Toward a Psychology of Being. Van Nostrand, New York

Maurer, R. (1996a) Put Resistance to Work for You. HRMagazine, 41

—-. (1996b) Transforming Resistance. Training and Development, Dec, 50, 12, 20–24

—-. (1996c) Using Resistance to Build Support for Change. Journal for Quality and Participation, Jun, 19, 3, 56–64

Meyer, J.P. and Allen, N.J. (1997) Commitment in the Workplace. Theory, Research, and Application. Sage, Thousand Oak, CA

Meyer, a.d.; Brooks, G.R. and Goes, J.B. (1990) Environmental Jolts and Industry Revolution: Organizational Responses to Discontinous Change. Strategic Management Journal, 11

Minet, F.; Parlier, M. and de Witte, S. (1994) La Competence : Mythe, Construction, ou Réalité? L'Harmattan, Paris

Mintzberg, H. (1973, 1980) The Nature of Managerial Work. Harper and Row

Morrison, K. (1995) Marx, Durkheim, Weber: Formation of Modern Social Thought. Sage, London

Nadler, D. and Tushman, M.L. (1977) A Diagnostic Model for Organizational Behavior. In Hackman, J.R.; Lawler, E.E. and Porter, L.W. (Eds.), Perspectives on Behavior in Organizations. McGraw-Hill, New York

Nalebuff, B.J. and Brandeburger, a.m. (1996) Co-opetition. HarperCollinsBusiness, London

Nelson, R.R. and Winter, S.G. (1982) An Evolutionary Theory of Economic Change. Belknap Press, ambridge MA

New, J.R. and Singer, D.D. (1983) Understanding Why People Reject New Ideas Helps to Convert Resistance into Acceptance. Industrial Engineering, 15, 5, 50–57

Nissani, M. (1997) Ten Cheers for Interdisciplinarity: The Case for Interdisciplinary Knowledge and Research. Social Science Journal, 34, 201–216

Noguera, F. (2003) L'Amenagement-Reduction du Temps de Travail Comme Outil de Gestion du Changement Organisationnel. Observer pour Agir, Revue de Recherche Appliquée, SupdeCo Amiens, vol. 2

Nonaka, I. and Takeuchi, H. (1995) The Knowledge Creating Company. Oxford University Press

Nonaka, I. (1991) The Knowledge-Creating Company. Harvard Business Review, Nov-Dec, p.96–104

Oldham, M. and Kleiner, B.H. (1990) Understanding the Nature and Use of Defense Mechanisms in Organizational Life. Journal of Managerial Psychology, 5, 5, i-iv

Olson, E.E. (1990) The Transcendent Function in Organizational Change. The Journal of Applied Behavioral Science, 26, 1, 69–81

O'Reilly, C. A. and Caldwell, D. (1980) Job Choice: The Impact of Intrinsic and Extrinsic Factors on Subsequent Satisfaction and Commitment. Journal of Applied Psychology, 1980, 65, 559–569

O'Reilly, C.A.; Chatman J., and Caldwell, D.F. (1991) People and Organizational Culture: Assessing Person-Organizational Fit. Academy of Management Journal, Sep, vol. 34 issue 3, 487–513

Ouchi, W.G. and Jaeger, A.M (1978) Type Z Organization: Stability in the Midst of Mobility. Academy of Management Review, Apr, vol. 3 issue 2, 305–315

Ouchi, W.G. and Johnson, J.B. (1978) Types of Organizational Control and Their Relationship to Emotional Well Being.

Administrative Science Quarterly, Jun, vol. 23 issue 2, 293–318

Owen, D.L.; Swift, T.A.; Humphrey, C. and Bowerman, M.C. (2000) The New Social Audits: Accountability, Managerial Capture or the Agenda of Social Champions? European Accounting Review, 9:1, 81–98

Penrose, E.T. (1959) The Theory of the Growth of the Firm. Blackwell, Oxford

Pepper, S.C. (1942) World Hypotheses. Berkley, University of California Press

Peters, T. and Waterman, R.H. (1982) In Search of Excellence. Harper and Row, New York

Pettigrew, a.m. (1985a) Contextualist Research: A Natural Way to Link Theory and Practice. In Lawler III, E.E.; Doing Research That Is Useful for Theory and Practice. Jossey-Bass, San Francisco, 222–249.

Pettigrew, a.m. (1985b) The Awakening Giant. Continuity and Change in ICI. Basil Blackwell, Oxford

Pettigrew, a.m. (1987) Context and Action in the Transformation of the Firm. Journal of Management Studies. vol. 24, n. 6, pp. 649–670

Pettigrew, a.m. (1990) Longitudinal Field Research on Change: Theory and Practice. Organization Science, vol. 1, n.3. pp 267–291

Pettigrew, a.m. and Whipp, R. (1992) Managing Change for Competitive Success. Basil Blackwell, Oxford,

Pfeffer, G. and Salancik, G.R. (1978) The External Control of Organization: A Resource Dependence Perspective. Harper and Row, New York

Piattelli Palmarini, M. (1995) L'Arte di Persuadere. Arnoldo Mondadori Editore, Milano

Polanyi, M. (1962) Personal Knowledge: Towards a Post-Critical Philosophy (Corrected ed.). University of Chicago Press, Chicago IL

Porter, L.W.; Steers, R.M.; Mowday, R.T. and Boulian, P.V. (1974) Organizational Commitment, Job Satisfaction and Turnover Among Psychiatric Technicians. Journal of Applied Psychology, 59, 603–609

Porter, L.W. and Lawler, E.E. (1968) What Job Attitudes Tell About Motivation. Harvard Business Review, Jan-Feb

Porter, M.E. (1996) What Is Strategy? Harvard Business Review, Nov-Dec, 61–78

Prahalad, C.K. and Hamel, G. (1994a) Competing for the Future. Harvard Business Review, Jul-Aug

—-. (1994b) Competing for the Future. Harvard Business School Press, Jul-Aug, vol 72 issue 4, 122–129

Quinn, J.B.; Anderson, P. and Finkelstein, S. (1996) Making The Most Of The Best. Harvard Business Review, Mar-Apr

Reagans, R. and Zuckerman, E.W. (2001) Networks, Diversity, and Productivity: The Social Capital of Corporate R and D Teams. Organization Science, 12, 502–517

Rogers, L. (2003) Lecture in the Course "Advanced Salesmanship Techniques." ISM PhD-DBA Seminar, 8–10 Dec

Rosenberg, M. (1999) Nonviolent Communication: A Language of Compassion. PuddlerDancer Press, Del Mar, CA

Salancik, G.R. (1977a) Commitment Is Too Easy. Organizational Dynamics. Summer

—-. (1977b) Commitment and the Control of Organizational Behavior and Belief. In Staw, B.M. and Salancik, G.R. (Eds.), New Directions in Organizational Behavior. St. Clair, Chicago, IL, 1–54

Schein, E. (1985) Organizational Culture and Leadership. Jossey-Bass

Schein, E. (2002) The Anxiety of Learning. Interview by Coutu, D.L. Harvard Business Review, Mar

Schneider, B.; Ashworth, S.D.; Higgs, A.C. and Carr, L. (1996) Design, Validity, and Use of Strategically Focused Employee Attitude Surveys. Personnel Psychology, 49, 695–705

Schneider, D.M. and Goldwasser, C. (1998) Be a Model Leader of Change. Management Review, 87, 3, 41–45

Schoemaker, P.J.H. (1995) Scenario Planning: A Tool for Strategic Thinking. Sloan Management Review, winter, 25–40

Schroeder, R. (1992) Max Weber and the Sociology of Culture. Sage, London

Schumpeter, J.A. (1967) The Theory of Economic Development. Oxford, Oxford University Press

Schutz, W.C. (1958) FIRO: A Three-Dimensional Theory of Interpersonal Behavior. Holt Rinehart and Wilson, New York

—-. (1966) The Interpersonal Underworld. Science and Behavior Books, Palo Alto, CA

—-. (1982) Profound Simplicity. Learning Concepts, San Diego

—-. (1984) The Truth Option: A Practical Technology for Human Affairs. Ten Speed Press

Senge, p.m.; Kleiner, A.; Roberts, Ch.; Ross, R.R. and Smith. B:J. (1994) The Fifth Discipline Fieldbook. Strategies and Tools for Building a Learning Organization. Doubleday, New York

Shell, G.R. (1999) Bargaining for Advantage. Negotiation Strategies for Reasonable People. Penguin, New York

Singh, V. and Vinnicombe, S. (2000) What Does "Commitment" Really Mean? Views of UK and Swedish Engineering Managers. Personnel Review, Vol. 29, No. 2, p.228–258

Skinner, A. (2001) Adam Smith, the Philosopher and the Porter. In Porta, P. L.; Scazzieri R. and Skinner, A. (Eds.), Knowledge, Social Institutions and the Division of Labor. Edward Elgar, Cheltenham, pp35–51

Smith, Ch. (2003) Managing Strategic Change. Handouts of the Course, MIB-ENPC, spring

Soparnot, R. (2003) Le Pilotage du Changement Organisationnel: de la Gestion du Changement à la Creation de Conditions Favorables. Observer pour Agir, Revue de Recherche Appliquée, Supdeco Amiens, vol 2

Sosik, J.J., Avolio, B.J., and Kahai, S.S. (1998) Inspiring Group Creativity: Comparing Anonymous and Identified Electronic Brainstorming. Small Group Research, 29, 3–11

Spender, J.C. (1996) Organizational Knowledge, Learning and Memory: Three Concepts in Search of a Theory. Journal of Organizational Change Management, vol. 9 n0.1, 63–78

Spender, J.C. (2004) Lecture in the Course "Knowledge Management." ISM PhD-DBA Seminar, 29–31 Jan

Steier, L.P. (1989) When Technology Meets People. Training and Development Journal, 43, 8, 27–29

—-. (1994) Choosing the Right Change Path. California Management Review, winter, p.29–51

—-. (2000) Pourquoi les Salaries Resistant-ils au Changement? In Le Changement. Harvard Business Review

Swift, T. (2001) Trust, Reputation and Corporate Accountability to Stakeholders. Business Ethics: an European Review, Jan, vol. 10 issue 1, p16–27

Tenenbaum, S. (2002) Nos Paysages Intérieurs. Ces Idées qui Nous Faconnent. InterEditions, Paris

Tessler, D.J. (1989) The Human Side of Change. Director, 43,3, 88–93

Tissier, D.(1988) Management Situationnel 1. Les Voies de l'Autonomie et de la Délégation. INSEP, Paris

Trice, H.M. and Beyer, J.M. (1993) The Cultures of Work Organizations. Prentice Hall, Englewood Cliffs, NJ

Trompenaars, F. and Hampden-Turner, Ch. (1997) Riding the Waves of Culture. Nicolas Brealey Publishing, London

Tsai, W. (2002) Social Structure of "Coopetition" Within a Multiunit Organization: Coordination, Competition, and Interorganizational Knowledge Sharing. Organization Science. Vol. 13, n. 2, Mar-Apr, 179–190

Tsui, A.S. (1990) A Multiple Constituency Model of Effectiveness: An empirical Examination at the Human Resource Subunit Level. Administrative Science Quarterly, 35, 458–483

Tuckman, A. (1995) Ideology, Quality and TQM. In Wilkinson, A. and Willmott, H. (eds.), Making Quality Critical. New Perspectives on Organizational Change. New York and London, Routledge, 54–81

Ury, W. (1991) Getting Past No. Randomhouse, London

Valverde, J. (2004) Lecture in the Course "Advanced Strategy." ISM PhD-DBA Seminar, 18–20 Jan

Van Krogh and Grant, cited by Choo and Bontis, 2002a

Vissers, G. and Dankbaar, B.; Creativity in Multidisciplinary New Product Development Teams. Creativity and Innovation Management, vol. 11 N.1 March 2002, 31–42

Vroom, V. (1964) Work and Motivation. Wiley, New York

Waddel, D. (1998) Resistance: A Constructive Tool for Change Management. Management Decision

Watzlawick, P. (1983) The Situation is Hopeless but Not Serious. The Pursuit of Unhappiness.

Watzlawick, P. (ed.) (1984) The Invented Reality: How do we know what we believe we know? Contributions to constructivism. Norton

Watzlawick, P.; Helmick Beavin, J. and Jackson, D.D. (1967) Pragmatics of Human Communication. A Study on Interactional Patterns, Pathologies, and Paradoxes. W.W. Norton and Company, New York

Watzlawick, P.; Weakland, J. and Fisch, R. (1974) Change. Principles of Problem Formation and Problem Resolution. W.W. Norton and Company, New York

Weber, M. (1978) Economy and Society (Roth, G. and Wittich, C., Eds.) University of California Press, Berkeley

Weick, K.E. (1979) The Social Psychology of Organization. Addison-Wesley, Reading (MA)

Weinbach, R.W. (1984) Implementing change: Insights and Strategies of the Supervisor. Social Work, 29, 3, 282–286

Whipp, R.; Rosenfeld, R. and Pettigrew, a.m. (1988) Understanding Strategic Change Processes: Some Preliminary Findings. In Pettigrew, a.m., The Management of Strategic Change. Basil Blackwell, Oxford, pp. 13–55

Wiener, Y. (1982) Commitment in Organizations: A Normative View. Academy of Management Review, 7, 418–428

Wilderom, C.p.m.; Glunk, U. and Maslowski, R. (2000) Organizational Culture as a Predictor of Organizational Performance. In Ashkanasy, N.M; Wilderom, C.p.m. and Peterson, M.F. (Eds.) Handbook of Organizational Culture and Climate. Sage, London, 193–210

Wiley, J.W. and Brooks, S.M. (2000) The High-Performance Organizational Climate. How Workers Describe Top-Performing Units. In Ashkanasy, N. M.; Wilderom, C.P.M and Peterson, M.F. (editors); Handbook of Organizational Culture and Climate. Sage Publications

Williams, L. C. (1993) The Congruence of People and Organizations: Healing Dysfunction from Inside Out. Quorum Books, Westport Connecticut.

—-. (1994) Organizational Violence; Creating a Prescription for Change. Quorum Books, Westport, Connecticut

. (1996) Business Decisions, Human Choices—Quorum Books, Westport Connecticut

—-. (2002), Creating the Congruent Workplace, Challenges for People and their Organizations, Quorum Books, Westport Connecticut.

Williamson, O.E. (1975) Markets and Hierarchies: Analysis and Antitrust Implications. Free Press, New York

Williamson, O.E. (1985) The Economic Institutions of Capitalism. Free Press, New York

Wilson, D.C. (1992) A Strategy of Change. Concept and Controversies in the Management of Change. Routledge, London

Witte, E. (1973) Organisation für Innovationsentscheidungen Das Promotoren-Modell. Göttingen

Woodman, R.W.; Sawyer, J.E. and Griffin, R.W. (1993) Toward a Theory of Organizational Creativity. Academy of Management Revie, 18, 293, 321

Woodward, D.G.; Edwards, P. and Birkin, F. (1996) Organizational Legitimacy and Stakeholder Information Provision; British Journal of Management, 7:4, 329–348

Zeigarnik, B. (1927) Das Behalten Erledigter und Unerlidigter Handlungen. Psychologische Forschung, 9, 1–85

TABLES AND DIAGRAMS

ENDNOTE

1 The "Oil Pricing" case developed at the Program on Negotiation at Harvard Business School is a very effective pedagogical tool to make people experiencing a complex version of the Prisoner's Dilemma.